Making Sense of Science

Making Sense of Science

Understanding the Social Study of Science

Steven Yearley

London • Thousand Oaks • New Delhi

First published 2005

SAGE Publications Ltd
1 Oliver's Yard
55 City Road
London EC1Y 1SP

SAGE Publications Inc.
2455 Teller Road
Thousand Oaks, California 91320

SAGE Publications India Pvt Ltd
B-42, Panchsheel Enclave
Post Box 4109
New Delhi 110 017

British Library Cataloguing in Publication data

A catalogue record for this book is available from the British Library

ISBN 0 8039 8691 2
ISBN 0 8039 8692 0

Library of Congress control number available

Typeset by C&M Digitals (P) Ltd., Chennai, India
Printed in Great Britain by The Cromwell Press Ltd, Trowbridge, Wiltshire

Contents

Introduction

SOCIOLOGY'S 'MISSING MASSES'

As I sit in my office in York making the final corrections to the text of this book, the British newspapers are full of a rival academic attraction under development in Yorkshire. Deep in a potash mine on the coast, about 100 kilometres north-east of my university's campus, scientists are refining their equipment in the Boulby Underground Laboratory for Dark Matter Research. They are working over 1000 metres below the earth's surface in the hope of screening out light and other familiar forms of radiation to such an extent that they will be able to detect the extremely subtle – indeed to date entirely undetected – minute entities known as WIMPs: weakly interacting massive particles.[1] These WIMPs may just contribute to solving a big problem for the physics community. Physicists today believe that the universe must contain more matter, more stuff, than is visible. If you count up all the stars, planets and black holes, the pulsars, comets, interstellar dust and so on, their mass and the gravitational force it ought to exert does not seem sufficient to keep the universe glued together. Physicists assume there must be 'dark matter' somewhere, stuff that has mass but which we somehow overlook. In Boulby they are looking for the missing masses.

The coincidence is that this book too is concerned with dark matter, in this case the masses missing from sociologists' accounts of contemporary society (see Latour, 1992). Society today is suffused with technologies and with insights and beliefs derived from science. Increasingly in modern cultures, citizens think about themselves and their own lives through the lenses of science. People label things about themselves – their habits or their appearance – as genetic (see Nelkin and Lindee, 1995: 14–18) or they worry about how to figure out the risks attaching to the food they feed their children or the inoculations to which children are exposed. With the spread of mobile phones, the once solitary pedestrian is now often in virtual contact with distant friends. People in the street have a new puzzle in trying to figure out whether passing pedestrians are talking to them or – via mobile phones – conversing with some absent person. There can therefore be no worthwhile sociology of present-day societies that does not have something to say about scientific ideas and the workings of science and technology.

One can put it more strongly than this. Scientific interpretations and technologies are in some sense 'actors' in current-day societies. Our world is 'peopled' not only with persons but also with computers, with viruses, with risks and with changing climates. This was made glaringly apparent in the US Presidential elections of 2000 when people's votes in several states could only be expressed through imperfect voting machines. It seems that the President was jointly 'chosen' by people and machines. Latour, whose work is discussed further in Chapter 4, has memorably described these other components of society as social science's dark matter. Machines and the other products of science and technology are sociology's missing masses because they go largely unnoticed by sociologists even while they bond the social world together. This dark matter holds social life together precisely because it enables us to lead our lives: voting machines (and now maybe email and text-messaging) allow us to cast our votes, while on-line travel timetables make it possible to schedule meetings. Yet we are still inclined to see the subject matter of sociology as only about people and institutions. If we want to understand how modern societies and cultures function, we need to take into account the workings of this dark matter too.

The key point, though in many ways an obvious one, is that the disciplines of science, technology and engineering are themselves attempts to explain how this societal dark matter operates. The study of microwave radiation helps us understand why mobile-phone signals can be received in some places and not others; atmospheric physics attempts to comprehend the basis of changing climates. The choice facing the social sciences is apparently stark. Once one recognises the importance of the missing masses to the basic task of sociology, should one simply delegate the explaining and understanding of this sociological dark matter to scientists and engineers, or should social scientists themselves try to say something about these missing masses? In the main, sociologists have overwhelmingly adopted the former alternative. A brief example can helpfully illustrate what is at stake here. Since the 1970s there has been intermittent talk of a worldwide energy crisis. The majority of people in the industrialised world are rampant consumers of energy. It is widely stated, though maybe not so widely reflected in official policy and private behaviour, that our demands for energy threaten to curtail the future prospects for social development. Most of our energy is derived from the sun though many industrialised nations also have large nuclear programmes.[2] People harvest some of this solar energy directly as well as gaining access to it more obliquely by using the winds or by burning timber; this wood-fuel derives from plants which employed the sun's energy to build the complex molecules of which they are composed. But the majority preference seems to be to use the energy which has been stored from these botanical processes in the past. Oil and coal deposits and the associated natural gas reservoirs result from the geological preservation of the remains of past plants and micro-organisms. Whichever sources of energy different societies favour, the total amount of energy available appears to be limited by the 'bottom line'. Societies cannot get hold of any more energy than the total afforded by the sun and the handful of other sources, unless of course chemists or physicists or other inventors devise new ways of generating energy.

In this sense, sociology appears condemned to operate within limits set by science, technology and engineering; society's 'dark matter' seems to be their business not ours. Of course, even if sociologists decided to leave the understanding of this dark matter entirely to the scientific community, that would not imply that social science had nothing to say about science and technology. Using the energy example we can clearly see that some countries have embraced nuclear power while others have been extremely wary; and this is not because the physics of nuclear energy are thought to vary from one state to another, from France to Norway for example. But it is equally possible to challenge the idea that there are objective, technical limits which sociologists simply have to accept. For example, recent sociological studies have examined how estimates of oil reserves are calculated, compiled and negotiated (see Bowden, 1985 and also Dennis, 1985). Estimates have changed a good deal over the last half century and at any one time there are usually competing estimates backed by seemingly well-qualified scientists, whether working for oil companies, prospecting consultancies or state geological and mineral survey bodies. Though no one denies that the reserves are finite, ideas about the extent of those reserves are subject to change. The energy limits facing contemporary society are not therefore simply physical constraints; there is an inescapable sociological dimension to the formation of ideas about those very limits. It seems that sociologists should not just hand over the study of society's dark matter to scientists and engineers because there is a sociological element to the missing masses themselves. Like the physicists in Boulby, we social scientists can try to study our own dark matter. The overall purpose of this book is to identify the best way to study that dark matter and to provide some examples which reveal the dividends for sociology of pursuing our missing masses ourselves. In a spirit of Yorkshire solidarity, I see this book's ambitions as paralleling those of the scientists in the potash mine.

THE SOCIOLOGICAL SIGNIFICANCE OF THE SCIENCE STUDIES

For most of the twentieth century sociology took a rather simple-minded view of science and its social role. It was assumed that science operated according to its own independent logic and that scientific thought was essentially beneficial or – at least – that it was a simple matter of weighing benefits and costs since, whatever its drawbacks, science did offer an accurate depiction of how the world was. This was also mirrored in some ways in the broader culture where science typically enjoyed high prestige and was commonly invoked in advertising campaigns and publicity slogans. Science was assumed to give rise to a greater understanding of the surrounding world, an ability to predict events and to control and manipulate parts of that nature. Despite wide, everyday acknowledgement of the importance of science to social change, science was comparatively neglected by many sociologists, perhaps because they lacked scientific training and felt overawed by the demands of science. Scientific institutions had tended to be neglected by the founders of the discipline also, though Max Weber had accorded scientific thought a significant role since it seemed to exemplify his idea of progressive rationalisation within modern capitalist societies.

Fundamental science, that is study of issues at the frontier of scientific understanding/ignorance, is often referred to as 'pure' science, and this terminology is informative as there was indeed something believed to be *pure* about science. Science represented knowledge of the world gained for its own sake. Rather like poetry, which follows its own standards of excellence, or the artistic accomplishments of leading painters, it was argued that science was best pursued according to its own devices. Though science might often turn out to be useful or 'applicable', this was not the prime objective or justification of science; its goal and ambition was the objective understanding of nature.

Consequently a distinction grew up between internal criteria (things which are properly to do with assessing science) and external criteria (things which are extrinsic to science, such as the economic benefits to be gained from new inventions or the political acceptability of the resulting ideas). When science was performed according to internal criteria alone, all would go well. Science ran into difficulties, however, when external criteria were allowed to intrude into the business of evaluating claims to scientific knowledge. The famous case always invoked here is that of the seventeenth-century church's opposition to Galileo's observational arguments in favour of the idea that the planets orbit the sun; the mainstream Christian church had always held that the earth was at the centre of creation and was not keen to see this idea overturned. Received wisdom is that the religious authorities came to be seen as having acted improperly by using external considerations to confound Galileo's internally robust arguments. This distinction between 'properly scientific' and 'external' criteria continues to be used both by liberals and conservatives (for example in debates over IQ and heredity, over the 'naturalness' of sex differences and the genetic basis or otherwise of homosexuality) to argue that if science said something was the case – however distasteful or unfashionable it might be – there was nothing to be done about it. Of late, this approach has mostly been used by conservatives to catch liberals out at their own game. Liberals supposedly subscribe to rational modes of argumentation and policy-making; if it turns out that science says that a propensity to crime is genetic, then typical liberal beliefs and policy recommendations are undermined. Conservatives, being less inclined to found their beliefs in pure rationality (because they are more inclined to give weight to tradition), are not quite so open to being caught out in this way.

A leading consequence of this view about the dangers of extrinsic considerations was that it came to be argued that science is best practised if it is left to the experts to regulate themselves. There was a guarantee that such science would produce valid knowledge if the scientists worked with internal criteria according to the 'scientific method'; this guarantee was underwritten by the system of publication in peer-reviewed journals which acted as a systematic check on quality. Science was a goose that could lay some really golden eggs, and politicians might like it to thrive for the sake of the golden eggs it could deliver (see Rip, 1982 on this analogy). But to rear it under circumstances where the eggs were all you looked for was self-defeating because the best eggs of all could sometimes arise unexpectedly from a most abstruse or unpromising area of research. Typically it is pointed out that obscure and 'pure' physics research led to the capacity for nuclear power (which, for a while at least, looked like a godsend) and that the 'chemical revolution' which made possible the nineteenth-century chemical industry was largely motivated by intellectual curiosity. In this way, scientists – particularly in the middle years of the twentieth century – won for themselves an extraordinary degree of freedom. As a profession, they were sponsored far more generously than the arts or any other aspect of high culture and yet were infrequently called to account for what they did, a privilege justified by the belief that only by this freedom could science work its magic. At the same time, this gave the scientific profession a very strong incentive to draw and police the boundaries around it. It became useful to be able to determine (and impose) limits to what was and what was not a science and to reward the 'in' activities with membership of scientific societies and other symbolic and material benefits (on this 'boundary work' see Gieryn, 1999).

The overall image here, therefore, is of a scientist as a contemplative figure, pondering the state of nature, primarily driven by curiosity and maybe the pursuit of recognition by his or her fellow scientists. The purity of this knowledge offered to put science into an unusual social role since no other knowledge could match its objectivity or detachment. Ideally the knowledge should reflect how the world is and thus the individual scientist would, in the end, only be a cipher. From a sociological point of view, it is a remarkable achievement for one group in society to have created a situation in which that group is believed to speak transparently about how the world is.

But this purity of science had another consequence too. It suggested that scientists were in a position to give disinterested advice. As pure scientists they had no vested interest except in the advance of knowledge. They would be able to speak the truth to powerful politicians and, if given the chance, could speak it unabashed. This became the warrant for increasing the social role of scientists in regulatory and advisory issues, whether giving evidence in the courtroom, advising on the environmental impacts of construction plans or providing advice to politicians and policy-makers.

For a couple of hundred years from the scientific revolution of the seventeenth century there was maybe something to this characterisation. Often scientists did science because it was of overwhelming interest to them; they received little payment to do it and often lived off a private income or made a living through employment as a minister of religion or a teacher. However, from the nineteenth century, science started to emerge and grow as a profession. Soon, companies began to take on scientists to work in their laboratories and pure science became increasingly unrepresentative of what most scientists, or typical scientists, do. And, once the majority of science is paid for by companies, by governments and by the military, it becomes far less easy to see science as disinterested. One has concerns about the extent to which the science can evade the influence of the commercial or political interests it is employed to serve.

Of course, it would never pay for scientists to be too partisan. A company wants to employ scientists to do competent research, not just to be yes-persons for the company. But to take the civil nuclear power industry as an example, although scientists are employed to do good nuclear design, the ones employed will, in all likelihood, be well disposed towards nuclear power. They will not take a detached view about the overall desirability or safety of the procedure. By the same token, scientists who work for environmental pressure groups will have to do technically competent science, but their fundamental orientation is most likely a given. Accordingly, in public disputes over technical matters the contest now most often resembles that between competing consultants rather than a process of contemplatively arriving at an informed verdict on the issue in question.

Thus, with the growth in the social, political and economic significance of scientists – most notably in the second half of the twentieth century – the automatic assumption of objectivity and impartiality wore off. And it has worn off not just those in the pay of companies, but off most scientists per se. For example, as science has grown as a profession, so the worry has increased that the scientific profession itself has vested interests to which scientists will attend. Scientists want better grants, more equipment and so on. They want to be taken more and more seriously by politicians and policy-makers. This has led to accusations that, for example, scientists have a vested interest in stressing the dangers of humanly caused global warming because it entails massive research projects, huge spending on science and an enlarged role for scientific advice-givers (see Chapter 11). Accordingly, it is sometimes alleged, scientists have jumped on the greenhouse warming bandwagon because it is in their material and professional interests. Such anxieties are commonly compounded by the way that scientific research has come to be associated with various kinds of societal ills, from the generation of toxic chemicals and air pollution to over-production and agricultural biotechnology. As will be seen in Chapter 8, there is an unfavourable association between science and social harm and a hint that risks are being imposed on society simply to allow scientists (and their commercial backers) to try out new experimental procedures. In recent years, Dolly the cloned sheep has probably become the favoured icon for this view of the scientific establishment.

The upshot of this is that the public position of science is now in some important respects almost reversed. Where once, only a quarter of a century ago, the primary worry among social commentators was that science would be regarded as disinterested when, in reality, it was not (see Habermas, 1971), the concern now is the opposite. Scientific judgements may be routinely doubted and associated with vested interests. To say that something is 'scientifically proven' is now as likely to be voiced ironically as literally. This anxiety has given rise to concern in the scientific community itself about the possible decline in the profession's public standing; in Britain this has led to an elaborate concern with the 'Public Understanding of Science' and to prizes and awards for efforts at promoting public understanding. Leaders of the scientific community seem to feel that both the public and the government are out of sympathy with the scientific estate, and various remedies have been proposed.

What this has meant is that the scientific profession has run into difficulties exactly where its golden-egg-laying characteristics might have been expected to win it most admiration. To choose a topical example, regulation is precisely the kind of area where one might anticipate that scientific knowledge would be able to do a useful public service. In principle we might all agree that we want to avoid the release of dangerous chemicals and the sale and adoption of inefficacious drugs. The way to regulate these things is to test their operation scientifically. But far from resolving the question once and for all, the recent experience of scientific testing – most notably in US courtrooms – is that it results in acrimonious and inconclusive conflicts among alleged experts.

The goal of this book is to analyse and understand, primarily for the benefit of a sociological audience, how these problems of science-in-society have arisen. My overall aim is to show that an appreciation of the difficulties confronting scientific authority and of current discontents with expertise, demands a sophisticated analysis of science, of sociology's dark matter. In this short introductory chapter I hope to have given the background to the central concerns of science studies. In my opinion, the sociology of science is essentially about two things: the sociology of the scientific community itself (in other words, the sociology of 'pure' science and other forms of research) and the sociology of that community's relationship to the rest of society. I consider that we need to look first at what the sociology of pure science looks like before we can understand the specific difficulties of science applied to policy, to law-making and to the analysis of politically sensitive questions. This will necessitate us looking at some historical examples as well as at the sociology of rather obscure and arcane science. It will be here that we find what is special about the recent achievements of science studies.

As will be seen, the sociological study of science has developed a great deal in the last 20 years. In institutional terms, the subject has made enormous advances, with major training programmes now existing in leading US, European and Pacific-Rim universities. More importantly, the academic achievements of this stream of work are now widely recognised and include some highly impressive studies. Furthermore, as we shall see, the sociology of science has had far-reaching effects on all manner of scholarly studies associated with the analysis of science, in particular the philosophy of science. The sole grounds for self-reproach among sociologists of science is, I suggest, that this institutional development and these scholarly achievements have had a relatively shallow impact on sociology and on social theory. Science studies has not, as yet, transformed sociology's understanding of its dark matter.

Accordingly, the plan of this book is in part to review these academic developments in a systematic way for a sociological audience (something which has been infrequently tackled to date[3]) and also to investigate and remedy the disregard for the sociology of science in social theory. The book's layout is straightforward: in Part I the central intellectual concerns of science studies are introduced. In Part II the leading 'schools' of the sociology of science are introduced, explained and examined critically. The schools selected for discussion not only indicate the range of theoretical and methodological positions represented in the recent history of science studies, they also exemplify rather different detailed interpretations of how the 'missing masses' are to be conceptualised. Following that, the longest, third Part of the text comprises a series of chapters dealing with what might be called 'applications' of science studies to particular topics of sociological concern. But to begin with, Chapter 1 starts with an assessment of what are commonly taken as the distinctive features of scientific knowledge, the features which make it seem like dark matter – like something occult – in the eyes of sociologists.

---

[1]WIMPs are 'massive' in the sense that they have some mass (albeit a very tiny amount) rather than being mass-less; they are far from massive in the everyday sense.

[2]Aside from the sun and nuclear sources, we also exploit some heat energy from the earth's interior and even extract some gravitational energy from the tides. Additionally, there are other ways of generating nuclear energy than the ones employed in power stations at present. Today's nuclear installations work on nuclear fission (energy released when atomic nuclei come apart); there is also nuclear fusion (where nuclei are fused together).

[3]Though see Lynch, 1993; Shapin, 1995 and, from a different perspective, Ward, 1996. For their enormously helpful comments on the text of this book can I please thank Barry Barnes, Eileen Crist and Darcy Binns.

PART I: THE CORE OF SCIENCE STUDIES

1 *Just What Makes Science Special?*

INTRODUCTION

As we saw in the introductory chapter, it is hard to deny that science is special. Science is the exemplar and the measure of knowledge in the contemporary industrialised world. Where religion once set the standard for sure knowledge, and logic was later elevated to the apex of human understanding, in the West science has now secured top position. Science tells us how the world operates. More than this, through its precision and mathematical form, science offers us a tight grasp on the workings of the world; and because of its energetic growth it offers to tell us ever more and more. These are all reasonable bases for thinking very well of science but, compared to the other exemplary forms of knowledge that have been revered in the past, a question hangs over the grounds for the special character of science.

Given the assumed greatness of God or gods, it was not hard to see why knowledge of the deity was assumed to be special and privileged. Societies devoted huge resources to institutions and projects associated with religious insights. If God, or His appointed spokespersons, said that something was the case, then it generally seemed safe to assume that it was indeed the case. Being God, He was not going to be wrong. In practice this did not always work out as straightforwardly as one might suppose. Even within the Christian tradition, for example, there have been frequent disputes over the status of religious knowledge. Alongside the idea of revealed truth coming from God, there has often been the accompanying thought that the true believer needs to have faith rather than the certainty that would come from direct evidence of God's intervention. Other

Christian thinkers turned the argument on its head, claiming that the marvel of the world is evidence of God's existence; some even tried to argue that one can get to the idea of God's existence through the business of reasoning alone. But the core conviction was that what we did know of God and of His views, on sin or redemption, say, we knew for sure. The specialness of the knowledge lay precisely in its certainty, its unquestionable nature. Logic seems to share God's transcendent quality. What was logical for Aristotle is logical for us. Logic may become more refined but things that were once logical do not become illogical overnight. The knowledge that logic offers us is exemplary because of its consistency and certainty. By contrast, scientific knowledge is often changeable and fallible. Even the best scientific ideas of a generation get overturned and most scientists agree that one of the exciting things about science is that it can alter so radically. It seems odd to venerate something so fickle.

Given that scientific knowledge is so changeable, its correctness cannot be the thing which makes it exceptional (unlike the idealised cases of religion and logic). Thus, when people have turned their thoughts to identifying the reasons for the exceptionalness of science they have come up with four kinds of ideas about how precisely its special character manifests itself. Within each of these four approaches there are important and fascinating internal differences but let us consider them in turn before making an assessment of attempts to specify what precisely is distinctive about science. It will be seen later that these fundamental approaches to the special character of science have spread far beyond philosophical and sociological discourse, for example to the worlds of the courts and of political debate, so that time spent on them now is a sound investment.

THE EMPIRICAL FOUNDATIONS OF SCIENCE

The empiricist version of science's special quality claims that scientific knowledge is special because it depends on systematic observation and measurement. Though no longer very popular with philosophers of science, this claim has enduring 'everyday' appeal. Empirical evidence is of course very important to science, as it is to art appreciation, train-spotting and picking the form of race horses. The important question is whether there is anything about the empirical foundations of science which sets it apart from other kinds of endeavour. There are several reasons for thinking not. In part, this is because observation in science is not just a matter of seeing clearly, even if scientific demonstrations are often set up to allow the audience to simply 'see' the truths of science. One sees many things on country walks but people need to be trained to 'see' the geology of the countryside, to observe the evidence of the last ice age in mounds of fractured rock, to see faults and discontinuities in the strata. Scientific observation requires interpretation rather than simply taking images in. But, the position of empiricists is further worsened by the realisation that contemporary science is overwhelmingly based on detection by machines and not by 'seeing' in any common-sense manner. Evidence of early life is found by using electron microscopes. The date of extinction events is observed by using isotopic dating techniques. Sub-atomic particles are observed using traces in cloud chambers. In a sense, each of these things is an observation but the seeing in each case depends on theoretical ideas embedded in, and taken for granted by, the techniques and equipment.

Although these points appear to be serious setbacks for advocates of the idea that science is special because it depends on observation, a worse problem arises when one examines the process by which scientific knowledge grows. In many cases, observations are at odds with other observations or with theoretical ideas. For example, at the end of the nineteenth century most geologists and many biologists believed that the earth was very old indeed, most likely many millions of years. They devised various methods for trying to get a figure for this age observationally, for example by trying to work out how much salt is added to the oceans annually and therefore how long the oceans must have been receiving eroded salts in order to become as salty as they are (on these efforts see Burchfield, 1990). At the same time, physicists were convinced that the sun could not conceivably be that old since it would have grown cold by now. Observation ran up against deduction from accepted theoretical beliefs but did not overturn it. There was always sufficient uncertainty surrounding any of the geologists' indirect 'observations' that they could be disregarded by at least many of the physicists. Other cases have come up with similar findings. Closer to the present, physicists are interested in radiation coming from the sun for what it may tell us about the reactions going on there. A leading component of that radiation is made up of solar neutrinos: curious virtually mass-less and non-electrically charged entities.[1] When scientists have sought to measure the flow of neutrinos these are claimed to be significantly more numerous than calculations from theory would lead one to expect. Someone who stressed the importance of observation might expect the scientific community to be overwhelmingly impressed by the measurements. But they are not (Pinch, 1980: 92). This is in part because the calculations are believed to be soundly based, but also because the measurements are technically so hard to carry out. It is hard to detect things which are almost mass-less. A complicated and costly experiment has to be devised to attempt to observe these neutrinos, and a whole chain of inferences needs to link the arrival of neutrinos with the ultimate print-out from the detector. The power of the observation is tied to the strength of each of those inferential steps, and there is no other source of firm evidence about the strength of those inferences. Worse still, the experimental set-up is so complex and costly that it cannot easily be checked or replicated.

The fourth difficulty with a reliance on observation is a point beloved of philosophers. Observations are of single things. Thus an astronomer interested in the development of stars has to base her or his claims on only certain observations. No astronomer can observe all the stars; there is not enough time. And, in any case, some past stars are now unobservable since we believe them to have ceased to exist already. Equally, the stellar-astronomer cannot observe future stars. Given that general statements cannot, for these reasons, be based strictly on observation, it turns out that there appears to be something undeniably theoretical even about general observation statements. It appears that a basis in observation falls short of justifying the enthusiast's claims for this being the key to the special nature of scientific knowledge. Scientific knowledge *is* based on observation but it cannot be exhaustively justified by observation alone. Apparently, therefore, science cannot be definitively separated from other cultural beliefs which also boast an observational basis.

THE SCIENTIFIC METHOD

Philosophers who wanted to retain the special character of science but who were aware of the flaws in an observation-based defence, looked for their salvation to other aspects of science. Most famously, Popper turned the last-mentioned difficulty on its head and tried to make it the characteristic strength of science. He realised that no amount of positive evidence for a generalisation really helped in the face of the enormity of potential negative observations. What one observed of the collapse of stars of a certain size, for instance, could be confirmed over and over by new observations, but these did not hold any weight against the enormous number of possibly disconfirming cases. However, the finding of one disconfirming case is, logically speaking, enough to demonstrate that one's generalisation is invalid. The sighting of just one black swan (in Popper's celebrated example) shows that the proposition 'all swans are white' is false. Popper accordingly shifted his emphasis from finding confirmatory evidence to the search for falsifications. Even if one could not prove the correctness of a generalisation by observation alone, he claimed that one could definitively falsify such a generalisation. Popper asserted that the distinctive character of science, therefore, was not observation (though observation is indispensable) but a commitment to falsificationism.

This analytical move paid dividends for Popper in several ways. First of all, it allowed him to separate science from various impostors. The truly scientific thinker formulates hypotheses and theories which are open to being falsified. With luck, they will not be. But they must in practice be open to the kind of testing which could falsify them. Charlatan theories, such as – in Popper's view – Freudian psychoanalytical theories and Marxism, provided themselves with alibis and excuses to forestall falsification. Falsificationism thus provided a criterion for demarcating properly scientific from pseudo-scientific theories. The second virtue was that Popper's approach sustained a key role for observation; scientific knowledge was founded in observation but it was not simply the accumulation of observational knowledge. Finally, Popper suggested that his understanding of science provided a methodological guideline for conducting science well: one should make bold conjectures and then be a ruthless falsifier.

There is a strong intuitive appeal to Popper's argument and his fundamental claim is commonly invoked by scientists (see Mulkay and Gilbert, 1981; see also the discussion in Chapter 10 dealing with science and the law). But, sadly for Popper, commentators have been able to point to many problems with his approach. For one thing the logic of falsification is not anything like as clear-cut as he would initially have had one believe. For example, to use the neutrino case again: the mis-match between the expected intensity of neutrinos and the measured amount appears to be a falsification of the theory. But it is quite possible that this has come about because the (fiendishly difficult) experiment is not functioning correctly. Thus, it requires a judgement to decide whether the disconfirmatory 'evidence' really is evidence or whether it is the result of experimental error. It could be that the experiment has been incompetently performed or that the inferential steps in the experiment themselves depend on an assumption which has been falsified. Early reports coming from Australia of the existence of a duck-billed platypus, an egg-laying mammal, were overwhelmingly discounted by Europe-based scientists essentially because it was assumed that it was much likelier that local observers were incompetent than that this apparent challenge to animal classificatory systems could be correct (Dugan, 1987). But, even if no practical doubt attaches to the observation and the assumptions on which it is predicated, one finds that scientists are commonly willing to tolerate anomalous findings. Sometimes they explain them away with after-the-event explanations; in other words they modify the theory in such a way as to accommodate the anomalous observation. On other occasions the scientific community appears simply to decide that it will shelve that objection for the time being. Kuhn, more historian than philosopher of science, noted how frequently this tolerance of exceptions occurred; for him it was a characteristic feature of the way that scientific thinking develops (1970: 18). He considered that anomalies were effectively stored up until a whole catalogue of anomalies could be invoked by challengers to the existing scientific orthodoxy.

Popper was sometimes more 'forgiving' of actual scientists' non-conformity to his principles than at other times (see Popper, 1972a: 33–59 for a rather forgiving version). But, it is clear that if it is at all common for scientists to use after-the-event explanations to account for anomalous findings or just to tolerate them, this makes science much closer to astrology than Popperians would ever be willing to concede and rather undermines the value of the demarcationary aspect of falsification. Moreover, as Popper unenthusiastically acknowledged, Darwin's theory appears to violate his falsificatory principles precisely because the notion of evolutionary benefit is so hard to confine. Darwinian field studies tend to accept that any apparently inexplicable feature of animal or plant design does have some evolutionary advantage. It's simply that it often takes a long while to work out what those advantages are. Even when possible advantages are identified, it is hard to work out a balance of costs and benefits. The peacock's lavish tail may win it more mates but tail-maintenance exacts a biological toll on the bird. Darwinians typically assume that the benefits must outweigh the costs; if not the tail would not be there. But this line of reasoning, present in field biology for over a century and thus apparently acceptable to the scientific community, threatens to become rather circular and thus (in a strict sense) unfalsifiable. Rather than treat any feature of plant or animal design as a potential falsifier of Darwinian theory, all anomalies are set aside until they can be fitted into the evolutionary paradigm.

In light of all these problems with his approach, Popper's supporters did not reject his theory but sought (in a rather non-falsificationist way) to amend it. The most ingenious revision was conducted by Lakatos (see 1978: 8–93). Lakatos proposed that the distinctiveness of science was revealed in the way scientists chose not between competing single theories but in their selection of groupings of theories; these he termed research programmes. He thereby introduced two key revisions to Popper's original scheme. He suggested first of all that it was not reasonable to give up a theory simply because it had been falsified; rather one adopted a new research programme only when its superiority over a preceding one had been shown. That is, one did not jettison one's theory in the light of counter-evidence; one only traded in a theoretical outlook for a superior one. Second, he accepted that it was inevitable that research programmes would produce after-the-event explanations for anomalous findings. Lakatos offered a spatial metaphor for his view of research programmes. They consist of a core of central theoretical commitments surrounded by a protective belt of more dispensable assertions. In his terms, it is rational for scientists to make changes in the protective belt to save the core from falsification. But in the end, the advance of science is still progressive and reasoned because there is a methodology for choosing between research programmes. A research programme can be pronounced to be degenerating when it is constantly responding to new evidence by having to make alterations in the protective belt. By contrast, a progressive research programme is where predictions derived from the theory are not being falsified by the evidence and where surprising predictions are being confirmed. Lakatos himself styles his approach sophisticated falsificationism, and its comparative sophistication is easy to see.[2]

However, there remain two problems with his approach. First, by acknowledging that it is reasonable to retain a theory even in the light of apparent falsifications and numerous anomalies, Lakatos robs himself of a clear methodological guideline. At some point, it appears, the reasonability of a theory/research programme must evaporate and it should become rational to switch out of the degenerating programme into a progressive one. But his theory cannot specify that moment. Popper's unsophisticated approach at least had the benefit of a definite cut-off point. It was irrational to adhere to a theory after it had been falsified. Lakatos can make no such claim. Secondly, Lakatos finds himself struggling to explain how his own theory matches the relevant empirical data. At times he appears to want to say that the best theory of the methodology of science is the one which makes most of the history of science look rational (1978: 121–38). One can understand his rationalist's fondness for that option but it is not at all clear that the version of 'the scientific method' which is most compatible with actual history is necessarily the one which would have been best.

Subsequent authors have taken up Lakatos' mantle – including his direct followers such as Zahar (1973), as well as Laudan (1977) and much more recently Kitcher (1993) – but each has had the difficulty of showing that their account of what is rational for scientists to do accords with what scientists actually do. Each has also struggled to identify the point at which it becomes rational to switch from one theory to another. In outline, the more elaborate and detailed the grounds for switching from one theory/research programme to another (the more sophisticated in Lakatos' sense), the harder it is to eliminate the scope for individual scientists to exercise judgement and thus to disagree with each other about which is the 'rational' path to take. These philosophers have usefully drawn attention to the constituents of scientific theories (the idea of a core and protective belt), and their exhaustive attempts to separate science from non-science turn out to have importance outside this apparently narrow, technical dispute (as will be seen in Chapter 10). But they have not achieved what they set out to do, to specify in close detail what makes science special.

LOOKING TO SCIENTIFIC CONDUCT

If the specialness or exceptional quality of science cannot be located in its method, perhaps it can be located in the social norms which govern conduct within the scientific community. This idea is associated with the earliest systematic analyses in the sociology of science and is particularly identified with the work of Merton who first elaborated this perspective in the 1940s. It was he who argued that there were norms generally accepted in the scientific community which 'possess a methodologic rationale but they are binding, not only because they are procedurally efficient, but because they are believed right and good. They are moral as well as technical prescriptions' (1973: 270). Such norms are thus important on two levels: they describe the normative atmosphere which in fact reigns within the community, and they operate in concert to make the scientific community effective in the production of sound scientific knowledge. As is well known, Merton principally suggested four candidate norms:

> *Universalism*: the belief that ideas should be evaluated according to impersonal criteria irrespective of their source. This norm should make considerations of gender or ethnic background or nationality, for example, unimportant to the assessment of contributions to science.
>
> *Communalism*: the principle that knowledge should be regarded as a common heritage and shared in the scientific community. Thus, scientists receive no payment for their publications in leading journals; indeed there is often a submission fee.
>
> *Disinterestedness*: the idea here is that scientists should not seek personal advancement in the scientific domain through questionable means nor should they advance vested interests through the medium of science. They should avoid 'eclipsing rivals through illicit means' (1973: 276) and ought not to promote the theories of their friends in the hope of back-scratching in return.
>
> *Organised scepticism*: the idea that scientists should not be credulous, not jump to conclusions, but weigh evidence in a considered manner.

The interesting and clever thing about these norms is that they govern the professional conduct of scientists and say little in detail about what one might think of as the 'scientific' aspects of their behaviour, the experimental protocols or what they choose to do in the lab or the field. But these norms none the less have implications for the growth of scientific ideas. For example, the norm of universalism suggests that scientists should, and will generally feel they should, take seriously all contributions to the scientific literature, whether authored by women or men, by gay or straight scientists, by those based in East Asia, Africa or the West. In this case one can quite readily see what Merton is getting at. Reports about, say, measurements of changing rainfall or ocean temperatures from Latin America – maybe as a result of climate change – could be just as important as those from Britain or France. Thus the norms allegedly describe how scientists conduct themselves and explain how such conduct collectively results in the growth of knowledge.

When this line of thinking was advanced by Merton and elaborated by his colleagues, they were able to offer various forms of support for the idea. At one level the proposal had an initial plausibility and seemed to describe how scientists in fact typically behave. Sharing ideas by being communitarian would seem to be inevitably beneficial to science. And the fact that scientists typically do publish their ideas freely and without payment seemed to suggest that there was something distinctive about the professional ethic of science. Secondly, Mertonian sociologists of science could point to cases where broader social trends which overrode the norms had caused disruption to science. To a considerable degree in Nazi Germany and to some extent in the Stalinist Soviet Union, scientific ideas were not treated in a universalistic manner. In one case, the ideas of Jewish scientists were held up to ridicule while, in the other, key ideas emanating from capitalist, imperialist countries were rejected. Mertonians argued that the pace of scientific and technical development slowed in both countries, at least relative to the comparatively universalistic USA, apparently demonstrating the utility of the norms for promoting scientific advancement. Finally, there were cases in the history of science where maverick scientists who had not adhered to the normatively prescribed patterns of behaviour had been subject to criticism. Perhaps most famously, the non-conforming eighteenth-century British chemist Priestley neglected to publish his path-breaking results. Mertonians offered evidence that other scientists responded with indignation to this improper conduct; other scientists' reactions seemed to support Merton's claim that communalism was experienced as a normative commitment. In summary, one can see how the norms should promote the growth of science; one can see that scientists commonly behave in accordance with the norms even at some cost to themselves (for example by publishing openly in journals even if there is a submission fee); one can see how deviation from the norms caused by the imposition of contrasting values (such as racist rejection of 'Jewish science') slows this growth; and one can find evidence that members of the scientific community respond with something like moral outrage to infractions of this normative code. The norms thus seem to govern how scientists conduct themselves and how they judge each other. In their community, scientists are rewarded for adhering to the norms and sanctioned for violating them. This social structure reproduces itself and speeds scientific development.

However, things have not worked out quite as straightforwardly as this view would suggest. For one thing, sociologists working within the Mertonian framework arrived at disconcerting findings. Mitroff conducted a study of scientists working on the origin and nature of the moon, using data from the US space programme. He conducted extensive interviews with members of the scientific community and looked for evidence of normative orientation in their remarks (1974: 27–46). He found statements supporting Merton's norms. But he also found support for contrasting behaviour which his respondents appeared to justify in puzzlingly similar terms. For example, scientists pointed out that, given the sheer amount of information potentially available, one had to limit the sources to which one paid attention. The work stemming from people located in well-known research groups could reasonably be given more attention than people who seemed to spring out of nowhere. Rather than acting universalistically, there were good functional grounds for doing exactly the opposite, being particularistic. Similarly, in order to get new ideas noticed, one had to champion one's innovative proposals. There were good grounds for not being disinterested. To give one's own novel suggestions a chance, one had to promote them above the opposition. In this way Mitroff argued that he had managed to find evidence for the existence of a corresponding set of counter-norms concerning conduct which was regarded as appropriate in the scientific community; he was also able to derive a functional justification for these counter-norms. Sharing and being communalistic are all very well, but there are sometimes grounds for being secretive. One wants to develop an idea to the state that it is reasonably robust before wasting other people's time by presenting it in the scientific literature. Similarly, if scientists took seriously all contributions to the literature and tried to check out the implications of every idea published, they would simply run out of time and scientific advance would grind to a halt.

Mitroff appears to argue that both sets of norms are operative simultaneously. There is a normative push towards universalism *and* towards particularism. He does not elaborate on how this state of affairs can be maintained. On the face of it, having both norms and counter-norms would seem to imply that there can be little normative control at all since more or less any course of action could be justified in the light of one set of norms or the other (Mulkay, 1980). The situation is not as grave as all that, however, since both the putative sets of norms focus on certain dimensions of scientific conduct. They could be read as indicating that, in the scientific community, there are particular sensitivities around issues of universality and of control over one's intellectual product.

This more relaxed interpretation of Mitroff's findings appears attractive in the light of the subsequent argument by Mulkay that Merton's supposed norms do not seem very well reinforced or rewarded in the scientific enterprise. Mulkay points out that there are some bits of scientific behaviour which are closely policed, rules for referencing in publications for example (1976: 641–3). Compared to those activities, conformity with the alleged Mertonian norms is barely policed at all. Rewards, in terms of prestigious jobs and research grants, flow to those renowned for good work and with long lists of highly cited publications. But these positive attributes seem to be the important things, not one's adherence to the behavioural norms. It is only an assumption (by Merton and his colleagues) that the two things – conformity to the norms and academic success – go together. But given Mitroff's findings and the fact that there appear to be so few institutional mechanisms for checking whether scientists do actually behave according to the norms, this assumption seems poorly supported by evidence.

These empirical difficulties for the Mertonian scheme suggest that the case for the four norms seemed convincing for several decades not because of the sociological accuracy of the norms alone but because, at a philosophical level, it appeared that these are the kinds of behavioural regularity that 'must' be enforced if science is to progress. Merton himself claimed that these norms are 'procedurally efficient'. For someone who approaches science as a straightforward empiricist, that is with an almost exclusive emphasis on observation, they would seem to be efficient behavioural characteristics. Of course, Mitroff already argued that, in practice, they might not be as efficient as all that. But from a post-Popperian perspective, their supposed efficiency looks even more suspect. Scientists have to decide which observations to count as 'real' observations and which to dismiss; beyond a certain point, being universalistic is a liability under these conditions. Similarly, as Lakatos' work indicated, scientists must judge whether a research programme is progressive or not and different scientists are likely to come to different conclusions. The injunction to be disinterested and to exercise organised scepticism will not be decisively helpful in making that judgement.

Merton's suggestion of founding the special character of science in the ethos of the scientific community was attractively novel. However, it appears that the evidence for the existence and institutionalisation of the norms is rather less robust than Mertonians had supposed. Worse still, it is not even clear that it would be good for the advance of science to have those norms institutionalised. Merton appears to be correct that certain dimensions of scientific conduct do have a moral or ethical salience to them, particularly certain issues which he lists under universalism (to do with equality of opportunity in science) and under communalism (concerning the ownership of scientific information); these matters arise again in connection with legal understandings of science in Chapter 10. But Mulkay's alternative interpretation, that the norms reflect a professional ideology developed by scientists to defend their independence and relative freedom from external scrutiny, seems to be as valid an analysis as that originally proposed by Merton (Mulkay, 1976).

SCIENTIFIC VALUES

If rules do not accomplish the task set out by rationalist authors of setting science apart from other forms of belief and if the scientific community is not distinguished by its normative ethos, another basis will have to be sought to justify the exceptionality of science. The other popular recourse for philosophical analysts has been to values. Kuhn, whose early work was mentioned above, sought to reduce or overcome the relativistic consequences of his earlier studies by suggesting that scientists consistently used a small number of key values for assessing the merits of rival scientific theories or research programmes. He proposed (1977: 322) that scientists prize highly the following five 'standard criteria for evaluating the adequacy of a theory': accuracy, consistency, scope, simplicity and fruitfulness. In assessing theories scientists will, on this view (1977: 321–2), evaluate them along the following dimensions:

1. the 'consequences deducible from a theory should be in demonstrated agreement with the results of existing experiments and observations';
2. the theory ought to be consistent internally and 'also with other currently accepted theories applicable to related aspects of nature';
3. the 'theory's consequences should extend far beyond the particular observations, laws, or subtheories it was initially designed to explain';
4. it ought to bring 'order to phenomena that in its absence would be individually isolated and, as a set, confused';
5. the 'theory should be fruitful of new research findings'.

Kuhn argues that scientists recognise that these features are desirable in scientific knowledge. In the language of advertising competitions, scientists use their 'skill and judgement' to assess the relative merits of contending theories or research programmes in the light of these values. The scientific community is the sole authority on the comparative standing of scientific ideas; the values which guide the growth of science are those which scientists collectively decide on. There is no other authority to which appeal can be made. As Kuhn states later in the same paragraph, these values 'provide *the shared basis* for theory choice' (1977: 322, emphasis added). These criteria are just a distillation of what scientists are found to do. One could, in an equivalent way, set out criteria encapsulating the activities of post-expressionist painters, successful romantic poets or leading exponents of dressage.

In Kuhn's statements, however, there remains an uncertainty about the precise nature, source and status of these criteria. For one thing, he accepts that the above is not a comprehensive listing; of these five values he says: 'I select five, not because they are exhaustive, but because they are individually important and collectively sufficiently varied to indicate what is at stake' (1977: 321). However, unless all the values could be listed it is hard to understand in what sense the values can be said to direct scientific decisions. Second, the status of the individual values is unclear. There is a tension between the view that they are simply generalisations about the values which scientists happen to honour – just as one might record the values recognised by followers of an artistic movement – and the suggestion that they have some intrinsic logic or that they derive from some transcendental standard. Third, as Bergström has helpfully pointed out in a thorough review of these arguments, such cognitive values are actually of different sorts, sorts which he labels as 'ultimate, evidential and strategic' (1996: 190).[3] While ultimate values directly reflect the underlying goal of science, evidential and strategic values act more as pointers towards that ultimate goal. Thus, the fifth criterion (fruitfulness) is not necessarily an ultimate value at all; rather one might select fruitful theories for strategic reasons (they allow the scientific community to identify new themes to work on) or on evidential grounds (one feels that a fruitful theory is likely to turn out also to be an accurate one). In Bergström's opinion Kuhn and related authors are unclear about exactly what makes the 'values' valuable.

Partly in response to these ambiguities Newton-Smith (1981) sought to provide a fuller defence of the use of values to preserve the rationality of science. His initial approach to the question differed from Kuhn's in that he began with a realist interpretation of science (for more on the meaning of 'realism', see the next section). Newton-Smith is cautious in his realism. He claims that science is distinguished from most other forms of knowledge because it tends to get truer as it goes along. Still, we cannot accept that our beliefs about the natural world at any particular time are the truth. Rather, we must accept the 'pessimistic induction' (1981: 14) that we will sooner or later abandon our current beliefs as untrue for, judging by the history of science, everything which we now believe true is likely to turn out to be false in some regard. We can, though, pick out criteria which have been used in assessing scientific ideas and which we have good reasons for thinking are linked to an increase in truthfulness or, as he terms it, verisimilitude. However, even though Newton-Smith is clearer about his identification of the ultimate justification for his values, several of the criteria he proposes are similar to those put forward by Kuhn. He lists a series of eight 'good-making features' of scientific theories (1981: 226–32). These are:

1. That a theory should 'preserve the observational successes of its predecessors'.
2. That a theory should be fertile in producing ideas for further inquiry.
3. That a theory should have a good track record to date.
4. That a theory should mesh with and support existing, neighbouring theories.
5. That theories should be 'smooth'; meaning that it should be possible to adjust the theory easily in the light of anomalies which are bound to emerge.
6. That a theory should be internally consistent.
7. That theories should be compatible 'with well-grounded metaphysical beliefs': that is, theories should accord with the same metaphysical assumptions as sustain the rest of science.
8. Although hesitant because of the ambiguity of this criterion, it is probably beneficial for theories to be simple.

What is significant about this list of criteria is not just the individual recommendations but the claim that the values each has a double justification. Newton-Smith asserts that they are both the criteria by which scientists do judge and criteria which can be shown to be rational for scientists to adopt in the light of the assumed goal of science, namely to become truer and truer. Thus, theories should be compatible with widely adopted metaphysical assumptions because it is very hard to see how science could be becoming more correct if major sections of it depended on conflicting metaphysics. If a new physical theory, for example, meant that while biology required the universe to be one way, physics entailed another ordering, that would be a retrograde movement.

Newton-Smith thus seeks to tackle Kuhn's problem head on: his theory is avowedly empirical and normative. It is an account of the values which scientists as a matter of fact generally do take into account and it is a demonstration of why scientists are right to honour those values. It is this latter aspect which would potentially allow Newton-Smith to claim that science is rational and that scientific knowledge is uniquely authoritative. But how satisfactory is this normative element? As Newton-Smith himself makes clear, none of the criteria is inviolable. On occasions some values may have to be subordinated to others. For example, a theory (T1) with a poor track record may be preferable to some other theory (T2) because of T1's assessment on the other values even though T2 has a better track record. In the great majority of scientific decisions, therefore, a judgement will have to be made about the merits of different theories' 'scores' on the eight values. And with these eight criteria to be taken into account the scores can be totted up in very many different ways. Just using the criteria will thus demand a huge exercise of judgement by scientists.

But the situation is even more complex than this for the criteria are not automatic in their application. Take criterion four for instance: meshing with and supporting neighbouring theories is a far from simple requirement. Which are the neighbouring theories? Looking back to the dispute outlined earlier about the age of the Earth, it would be evident for supporters of the geological position that the study of the growth of biological diversity was a field neighbouring the study of the Earth's age. For physicists, however, the study of biological phenomena would be only very remotely connected to the issue of the probable antiquity of the Earth. But even if neighbours could be uncontentiously identified it would still be unclear how to evaluate the degree of support given to those neighbouring theories. Is it better to lend a great deal of support to a few neighbouring theories or to lend some support to a lot of neighbours? When viewed in this way it appears that Newton-Smith's approach is subject to the same practical limitations as those Kuhn (1977: 324) admitted for his own since:

> When scientists must choose between competing theories, two men fully committed to the same list of criteria for choice may nevertheless reach different conclusions. ... With respect to divergences of this sort, no set of choice criteria yet proposed is of any use. One can explain, as the historian characteristically does, why particular men made particular choices at particular times. But for that purpose one must go beyond the list of shared criteria to characteristics of the individuals who make the choice.

Later on in the same text Kuhn reinforces this point, acknowledging that 'little knowledge of history is required to suggest that both the application of these values and, more obviously, the relative weights attached to them have varied markedly with time and also with the field of [science in which they are applied]' (1977: 335).

Newton-Smith took Kuhn to task (1981: 122–4) for not rooting his proposed values in the rational requirements of science. For him, Kuhn was making too weak a case for science by implying that the five Kuhnian values were just a statement of how scientists happened to conduct themselves. As a realist, Newton-Smith cannot allow that science comprises a set of values or criteria one can choose to follow or not. The values must not just be a convention; they have to be the real values for getting on best in describing the world. But, as we have seen, the practical normative force of the proposed eight values is much less than Newton-Smith would seem to require.

One may accept that there is a certain plausibility to the values. They may describe the kinds of considerations which scientists appear to have in mind when selecting theories; they may even strike us as the kind of consideration which scientists ought to have in mind. But, unless we have good reasons for thinking that the values direct scientific choices in a strong sense, this normative force is of limited consequence. Providing a list of values which scientists should honour but which, in practice, does not constrain scientific choice at all closely, does rather little to revitalise the authority of any specific scientific judgements. Newton-Smith supplies us with general grounds for thinking that science as a whole is a reasoned undertaking but he does not reassure us that any particular scientific judgement could not reasonably have come out differently. By listing his suggested criteria in a chapter entitled 'Scientific Method', Newton-Smith might be seen as implying that the criteria can be used as something like a recipe for demonstrating the exceptionalness of scientific progress. It should now be clear that they cannot serve in this capacity.

REALISM

Though philosophers such as Popper, Kitcher and Kuhn have engaged in their different ways with the issue of how it is that science secures its position of special authority, a rather different line of argument has been developed by other philosophical analysts (including, to some degree, Newton-Smith). Their position is commonly termed realism. They are much less concerned with the mechanics of scientific advancement than with considering the status of the entities (particularly theoretical entities such as scientific 'laws') posited by scientists. The realist position maintains that the things disclosed by science are among the real constituents and the real mechanisms of the natural world; they are – in the philosophical cliché – the furniture of the universe. Given that realists believe that science tells us about the real fabric of the world, it is in some sense quite unnecessary to worry about how exactly science manages to be progressive. For realists, the important thing is that the scientific endeavour tells us how the world is; the fact that it does this is far more important than the secondary issue of how it does it. And even if we cannot at present specify in detail how science does it, that will not stop realists claiming both that it does do it and that we know that it does.

Some philosophers had seen the 'how' question as the route to demonstrating the superiority of science. But realists typically use different arguments. They commonly concentrate on figuring out the way the world must be if we humans are to have knowledge of it. That is to say, realists use transcendental arguments to work out what the fact of human knowledge tells us about the relationship between humans and the natural world. Bhaskar has stated this as clearly as any of the realists:

> It is not necessary that science occurs. But given that it does, it is necessary that the world is a certain way. It is contingent that the world is such that science is possible. And, given that it is possible, it is contingent upon the satisfaction of certain social conditions that science in fact occurs. But given that science does or could occur, the world *must* be a certain way. Thus, the transcendental realist asserts, that the world is structured and differentiated can be established by philosophical argument; though the particular structures it contains and the ways in which it is differentiated are matters for substantive scientific investigation. (1978: 29 original emphasis)

The realist argues that, for science to exist, the world must have certain properties or characteristics. And humans, as part of that world, must have certain characteristics as well. It is important to see that the claim here is not a narrowly factual or empirical one. Realist philosophers move from the fact of science's existence to deduce what the world must be like, in general terms, for science to be possible at all. Their main appeal is to reason – to pure thinking – not to detailed claims about the actual behaviours of scientists.

Given this orientation, it goes almost without saying that realists are not primarily interested in trying to demonstrate how it is that scientists' activities or procedures are able to produce a special kind of knowledge. They tend to take it as a given that science is successful, and then aim to work out what this implies about the nature of the world and our relationship to it. For example, Bhaskar's argument is not intended to convince people who believe that science is unsuccessful. He is trying to show that analysts of science who think – as an extreme follower of Popper might – that science is composed only of competing, alternative hypotheses are mistaken. He proposes that the practice of science makes no sense without two separate presuppositions. The first is that the objects of scientific knowledge are independent of the activity of science itself. The second is that scientific knowledge can only be produced by a community of knowers; it is not the spontaneous product of individual observers' perceptions. Thus, for example, his retort to the Popperian is that 'To be fallibilist about knowledge is to be realist about the world' (1978: 43). The very idea of falsifying hypotheses makes no sense unless one assumes that there is an independent natural world with the capability of falsifying our proposals; to be a Popperian is thus (says Bhaskar) implicitly to endorse realism.

For this reason, the realist's best argument is that, if we think about it, the very business of engaging in science presupposes realist assumptions. They believe that any other position is untenable; scientists' actions would be at odds with those alternative claims. Even if people deny that they are realists, their very conceptualisation of science belies their words. Realists' claims for the special character of science are a consequence of this argument; science is special because it tells us about the real causal structures of the world. Quite reasonably, realists view this as no mean feat and thus as considerable evidence of special-ness.

As indicated in the quote above, realists such as Bhaskar acknowledge that their philosophical arguments are limited to establishing that 'the world is structured and differentiated' but can tell us nothing substantive about how the world is, since that is the business of science. Given this limited objective, one might wonder what use realists suppose their arguments are. The prime answer is two-fold. In part, their arguments are intended to put a stop to misunderstandings about what science and the scientific community must be like; Bhaskar believes that Popper and Kuhn and many others are barking up the wrong tree, thus wasting time and perpetuating mistakes about the status of science's discoveries. Secondly, he appears to believe that the practice of knowledge-making sometimes goes wrong because it is allied to a false philosophy. In Bhaskar's case, he wishes to reform social science (in a neo-Marxist direction) and wishes to outlaw other schools of social-scientific thinking by showing that those schools are philosophically untenable.

Proponents of the other arguments I have reviewed may or may not explicitly identify themselves as realists; of course, Bhaskar would wish to claim them all as realists, at least at an implicit level. Thus Newton-Smith claims to be a temperate realist and bolstered his claims about the good-making features of science with arguments from realism. He argues that the conceptual values are compatible with transcendental arguments about how scientific knowledge and the real world must be. Popper was apparently much more impressed with the fallibility of science. Too much realism about any existing conception was likely to be misplaced since scientific development entails a constant challenging and overthrow of existing ideas. In Newton-Smith's words, Popper was struck by the pessimistic induction that all current science is likely to turn out to be wrong. In this regard, it should also be noted that realists, while typically realist about the empirical and experimental sciences, are also often realist about arithmetic, geometry and other forms of abstract knowledge too, as will be seen in the next chapter.

The kind of arguments advanced by Bhaskar and other realists have clearly exercised a strong appeal, but in important ways they are both too strong and too weak. They are too weak in the sense that, even if one accepted them, that acceptance would very often have few consequences. In a controversy in the scientific community, realism will not generally help one decide which position to favour since, as Bhaskar acknowledges above, 'the particular structures [the world] contains and the ways in which it is differentiated are matters for substantive scientific investigation'. Similarly, realism will not typically help policy-makers decide which scientists' advice to heed or help a court decide to which expert witness it should pay most attention. At the same time, the argument is too strong because it appears to use transcendental arguments to demonstrate the existence of a real world when the only thing knowable about this world is that it is real. It seems to solve the problem of the exceptional character of science but only does so by inferring the existence of a real world about which we can know nothing except those things which scientists have already told us. In that sense, it is a little like transcendental arguments for the existence of God: arguments that purport to tell us that God exists but which leave everything else important about God to the sources which previously informed us. In this way, the argument seems perilously close to circularity. This issue of the status of realist arguments will be considered again in Chapter 2.

CONCLUDING DISCUSSION

This chapter has been concerned with trying to pin down precisely the source of science's exceptionalism. If scientific knowledge is to stand apart from other forms of knowledge in contemporary society then one would presume there would be an identifiable basis for that distinctiveness. Analysts of science have identified four main routes for attaining this Grail. However, though each of these approaches is partly persuasive, none achieves its initial goal. The only philosophical approach (realism) which comes close to making science stand out and be truly exceptional pulls off this trick by claiming that the practice of science *necessarily* implies that the world is real and that science gives us access to that real world. Realism insists that science is exceptional but the only evidence is the existence of science itself.

While the reviews conducted in this chapter largely point to dead-ends if one's interest is in proving how exactly science is exceptional, that does not mean that the approaches have been futile. For one thing, many of the arguments considered here turn out to be important later on when we come to analyse the standing of science in court or the role played by scientists in advising on policy. On top of this, the analysts whose work has been considered have made useful contributions to the study of science, even if they haven't achieved all they set out to do. Popper's observation about the importance of falsification and falsifiability will crop up again several times. Lakatos' distinction between the central core and the protective belt provides an important way of describing the structure of many scientific theories. Merton's emphasis on universalism has continued to play a key role in the study of controversies involving science while the kinds of concern raised by realists turn out, perhaps surprisingly, to be critical to many schools in the sociology of science including the ethnomethodologists. Finally, it will be seen that the cognitive values emphasised by Kuhn and Newton-Smith mirror in an interesting way the manner in which a sociological analysis of science has most successfully been developed. It is to that programme of studies in the sociology of science that we turn in the next chapter.

---

[1]Relating back to the introduction and my discussion of the search for WIMPs, it should be pointed out that neutrinos themselves are thought by some to be part of the dark matter. But for neutrinos to make up any substantial proportion of the missing masses, they would have to be found to have more mass than is generally reckoned to be the case. It is possible that there are different kinds of neutrinos, some being WIMPy, others not.

[2]Even though Popper was sometimes less naïve a falsificationist than at other times, he was never this sophisticated.

[3]My thanks to Alan Weir of the School of Philosophical Studies at Queen's University Belfast for alerting me to Bergström's analysis.

2 Framing Commitments: The Strong Programme and the Empirical Programme of Relativism

INTRODUCTION: THE PROGRAMMATICS OF THE STRONG PROGRAMME

The symbolic heart of the new sociology of science is Bloor's 'Strong Programme'. The Strong Programme aims to wrest the study of science from the grasp of philosophers, or at least from the kinds of philosophies reviewed in the last chapter. It does this by denying the working assumption of the authors considered in Chapter 1; the Strong Programme refuses to take for granted – indeed tends in key respects to deny – the exceptionality of science. In his book *Knowledge and Social Imagery*, first published in 1976, Bloor set out an agenda for the sociology of scientific knowledge (hereafter SSK) which has become the talismanic point of reference for many later works. The irony here is two-fold: in 1976 Bloor had very few detailed case studies or other forms of empirical work to support his claims; moreover, very few subsequent practitioners of SSK appear to follow Bloor's tenets in specific detail. Yet, the book's exemplary status is reflected in repeated citations and in its re-publication in 1991, with a new 'Afterword'. On its first outing it won notoriety for espousing a 'Strong' approach to the sociology of knowledge. The sociology of knowledge, says Bloor, examines the interaction between whatever counts as knowledge in a particular culture and the social characteristics of that culture. It is easy to imagine a sociology of knowledge about astrology, about views on design and fashion, about artistic styles or about racial stereotypes. But certain types of knowledge might be thought to be immune to such inquiry. The supposedly universal truths of logic might be expected to have impressed themselves equally on all cultures. Sociologists would be wasting their time looking for social influences on ideas about logic.

What was strong about the Strong Programme was its insistence that social science should treat all kinds of knowledge equally. The social scientist should adopt the same 'impartial' approach to explaining people's beliefs about science or mathematics as he or she would adopt for analysing beliefs about religion or political ideology. Even more radically, this equality of treatment should be extended to the explanation of beliefs which come to be regarded as true or as false. As Bloor puts it, the sociology of knowledge 'would be symmetrical in its style of explanation. The same types of cause would explain, say, true and false beliefs' (1991: 7). Bloor's programme contains two other tenets in addition to impartiality and symmetry; he terms them causality and reflexivity. By insisting that the sociology of scientific knowledge should be causal, Bloor means that SSK should be 'concerned with the conditions which bring about belief or states of knowledge' (1991: 7). In using the much-debated and philosophically controversial term 'cause', Bloor appears only to be meaning that the sociology of knowledge should aim to explain how knowledge comes to be taken for what it is in any given society; he is not explicit about what form that sociological explanation should adopt. By reflexivity Bloor means that 'In principle its pattern of explanation would have to be applicable to sociology itself' (1991: 7). Bloor took little subsequent interest in this tenet (see Ashmore, 1989: 20 and the discussion of reflexivity in Chapter 7).

Stating Bloor's thesis in this bare fashion, it is easy to imagine the indignation it prompted. His support for symmetry and impartiality appeared to put all knowledge on the same footing and, implicitly, to suggest that today's science was no better than yesterday's, or – indeed – than witchcraft. Much of the controversy which followed the book's publication was conducted at an hysterical level. Philosophers, natural scientists, anthropologists and psychologists joined the dozens of sociologists who, at conferences and in reviews, unleashed their 'ultimate refutations' of Bloor's work. Bloor even found himself sharing lecture platforms with parapsychologists and other cognitive deviants, invited to philosophers' conferences as an epistemological freak-show. Yet he evaded their knock-down arguments, many of which – as he deftly showed in his Afterword (1991: 163–5) – were formulated with the benefit of only passing acquaintance with his book. In assessing the importance of his arguments, therefore, one needs to set aside the controversialising as much as possible. At base, Bloor is arguing that all knowledge should be studied with the same tools and with the same explanatory end in mind, not – as many critics assumed – that all knowledge is therefore the same. The tools he wants us to use are naturalistic ones; that is, he wants us to explain the bodies of knowledge which societies develop in terms of this-worldly, empirical factors. These explanatory factors will primarily lie in the biology, psychology, sociology and politics of those societies. The important thing for Bloor is that the explanation of knowledge, whatever way that explanation turns out in any given case, should be in terms of naturalistic causes.

If Bloor is not trying to assault all knowledge and make it out to be nothing but arbitrary social convention, what is his target? The answer is: non-naturalistic forms of explanation. Bloor's chief campaign is against 'explanations' of knowledge which are couched in terms of entities and standards which transcend empirical reality. He is an unremitting opponent of all such accounts and appeals. To clarify what is at stake here, let us take the example of ethics. In arguments about what to do or how to live, people commonly make reference to ethical principles. But when ethicists attempt to work out what the status of these principles is they run into a dilemma. Speaking very loosely, the ethicists' problem can be summarised as follows. Ethical principles are ideas. But if they are just individuals' ideas this makes them too subjective and cannot help us figure out what it is that is good about, say, freedom or justice. On the other hand, it is hard to envisage what there is 'out there' with which these ideas can correspond and which can thus differentiate good ethical ideas from bad ones. All in all, therefore, it is difficult to defend ethical realism. Ethicists have shown great ingenuity in devising intermediate but somehow transcendental states in which ethical principles supposedly exist. But one runs into difficulties as soon as one tries to use these intermediate states to explain people's ethical conduct. How do these transcendental things interact with the everyday causes which propel human activity? In the technical philosophical literature, and with enchanting disregard for everyday sensibilities, such properties are referred to as queer. The label 'queer' reflects the fact that philosophers provisionally agree that ethical properties are neither simply subjective nor straightforwardly objective; loath to give up on the existence of these properties, philosophers invent another kind of existence for them, a third way. It is part of the oddness of these queer properties that people can somehow sense or 'see' them even though queer properties are (by definition) not the same as everyday empirical attributes.

Bloor argues the same way about the mathematical approach to number. Reviewing the arguments of the late-nineteenth-century mathematician Frege, Bloor proceeds more or less as above. Frege appreciated that number is not merely a psychological phenomenon; the notion of number is more objective than it would be if it were just individuals' ideas about number. But neither is it directly material since, though one can count material things such as apples or oranges, it is possible in mathematics to work entirely with abstract numbers. Hence, Frege concluded, number has to have a different, a third status. It is at this point that Bloor's argument becomes most distinctive. Frege wanted to invoke some non-naturalistic cause for, or explanation of, our beliefs about such matters; he described this third type as 'objects of Reason' (Bloor, 1991: 96). Bloor argues that such a manoeuvre is illegitimate since there is no plausible account of what this third level of existence is nor how our everyday minds can have access to this third, 'queer' realm. Still, Bloor does not want to deny that we experience such concepts as having a third status. Instead he asserts that the only candidate is the social: 'that very special, third status between the physical and the psychological belongs, and only belongs, to what is social' (1991: 97). This pursuit of some 'third way' is quite general. For example, late in his career Popper coined the term 'World Three' to refer to – as he puts it – an 'autonomous' world apart from the worlds of physical things and of individuals' thoughts about them. He proposed the existence of 'three ontologically distinct sub-worlds' (1972b: 154). By arguing that this experience of a third world can only stem from the social, Bloor restores the symmetry between the causes of belief and avoids the need for multiple worlds as invented by Popper. According to the Strong Programme, our beliefs are brought about by some combination of psychological, physical and social factors. 'Queer' properties, which appear at first sight to be neither physical nor psychological, can still be understood in naturalistic terms if their origin is seen as located in patterns of social compulsion. The relative contributions of the explanatory factors in any given case are less important to Bloor than the fact that all three are equally naturalistic.

The suggestion that there should be a thoroughly naturalistic approach to knowledge is not peculiar to Bloor. In the preceding decade Quine had argued for the naturalising of epistemology, meaning that psychological investigations would resolve such epistemological queries as could sensibly be answered. 'Epistemology, or something like it, simply falls into place as a chapter of psychology and hence of natural science' he had asserted (1969: 82). In making such statements Quine's philosophical impudence was every bit as great as Bloor's. But what excused Quine from much of the hostility directed at Bloor was the fact that a psychological naturalisation can appear quite unthreatening. As I indicated in Chapter 1, we have long ago grown used to talking about knowledge through the metaphor of perception. Knowing clearly is – according to this analogy – like seeing plainly. Thus, for Quine to say that the study of human knowledge can only be the study of psychology does not, at first sight, jeopardise the quality of knowledge, especially empirical knowledge. But Bloor takes this line of reasoning forward in two directions. First, he points out the inadequacy of biology and psychology for any reasonably convincing naturalistic account of how knowledge develops. People do not form ideas, refine concepts, or develop theories as individuals; even Bhaskar accepted that much. Rather, knowledge results from the interactions of humans; consequently this sociological or cultural dimension must be reflected in any naturalistic account of knowledge (1991: 168–9). Bloor can present this argument quite innocently as the completion of the naturalistic turn in our understanding of knowledge. But he then goes on to point out that the causal factors of a sociological kind are liable to vary from one culture to another. There are therefore likely to be alternative traditions of knowledge about logic, mathematics or science, all equally brought about by naturalistic causes. The comparison with Quine is again apt. Quine's argument makes it senseless to try to get to a pure epistemology by stripping away the psychological components; without our psychologies there would be no knowledge, so the objective is self-defeating. Bloor argues that knowledge is always social; it cannot be made better by eliminating the social.

BLOOR ON MATHEMATICS AND ON NATURAL SCIENCE

Although he does not emphasise this point himself, Bloor is unusual among recent sociologists of knowledge in dealing with mathematics as well as with scientific knowledge. Since the first publication of his book there has been a vast growth in the field of science studies. Bloor's concern with mathematics has been less well matched by a growth in scholarly publications. Yet it is precisely here that he is at his most radical and challenging. It is common to find a special third status claimed for mathematics and logic. Thus, mathematical truths appear to have a transcendental character, to partake of some ideal quality while they are also capable of being reflected in our minds. Bloor offers a materialist challenge to this notion. He adapts John Stuart Mill's empiricist view of mathematics: that it is grounded in generalisations learned from manipulations of the physical world. But in Bloor's account the sense of compulsion associated with the third status is retained as well; Bloor wants to be true to the 'feeling ... that some Reality [that is, some transcendental reality] is needed to account for mathematics. On the present theory this feeling is justifiable and explicable. Part of that reality is the world of physical objects and part of it is society' (1991: 105). The development of mathematical knowledge is open to sociological investigation precisely because the other-worldly 'objectivity' of mathematics arises from social conventions. Bloor goes on to demonstrate that these conventions are social/cultural by illustrating negotiations over, and changes in, mathematical standards and proofs. Thus, at one point in time it was regarded as valid to divide a curve conceptually into a vast number of infinitesimally small straight lines in order to make proofs about geometric figures or to definitively solve equations; at other times such a manoeuvre was regarded as illegitimate. The 'truth' about infinitesimals does not legislate whether such procedures are acceptable or not. There are only disputes over whether this tradition, this convention – rather like the convention of pointillism in painting – is acceptable or not.

At first sight, it appears odd that Bloor should concentrate his argumentative forces against theorists of World Three. But, his aim is to show that this type of view is actually the only full-blown alternative to his naturalism. The only reason not to be symmetrical in our treatment of truth and falsity would be if believing in true beliefs were somehow different from believing in false ones (1991: 178–9). If there were something distinctive about true beliefs – if they were a different colour or smelt different – then one could imagine how believing true things would differ from believing false ones. But there clearly is no difference. People can be as convinced of, as passionate or as scrupulous about beliefs which they later come to see as false, or which others see as false, as about beliefs which are regarded as true. We use processes of argument, of experimentation, of observation to work out which beliefs are true and which are not. The ascription of truth is the outcome of these processes. In everyday usage, we say that the truth of the belief was the cause of the outcome. Bloor's point is that this is only correct in a circular sense. Without the processes, experiments, observations and so on, we would never have known what the truth was. The carrying out of those processes and so on explains why we come to believe what we do. There is nothing additional about those beliefs that marks them as 'true'. Had there been any such markers, we would not have needed to do the experiments and observations in the first place. The process must remain inscrutable, as Quine quippingly remarked, because there is nothing there to scrute. We further dignify the successful ideas with the acclamation of truth, but – though it is very common to talk in this way – it is no more than self-congratulation. The only way to argue that there could be some special way of recognising true beliefs would be to make those truths extraordinary in some way, thus making truth too into a queer property.[1]

This, then, is the reason why Bloor treats mathematics and logic as important test cases. Few philosophers want to join Popper in elevating correct empirical beliefs into some parallel third realm. But logic and mathematics have often seemed to philosophical analysts to have more than an empirical justification. If Bloor can convince himself and his readers that even mathematics and logic have naturalistic explanations, then empirical natural science will easily follow.

However, sociologists have tended to be more interested in studies of the empirical sciences than in the abstract realms of geometry and algebra. In particular, they have been interested in those areas of science where scientific developments are interlaced with economic and political debate, as with aspects of Darwinism, eugenics, machine intelligence and so on. The curious thing is that Bloor appears less radical here than in relation to mathematics. When it comes to natural science, Bloor's radicalism is not so disturbing or outstanding. Primarily this is because natural science is avowedly empirical; it does not usually share the Fregean, transcendental ambitions of mathematics and logic. Consequently, Bloor's claim that our sense of the 'Reality' of our knowledge is partly due to the world (including our psychology) and partly due to society is not such shocking news when it comes to biology or astronomy. A materialist account of scientific knowledge was always going to be less provocative than a materialist account of number or of logic.

Indeed, proponents of conventional views of science can draw some consolation from the fact that, as an advocate of naturalistic explanations, Bloor appears committed to giving the physical world an explanatory role in the development of scientific knowledge. The 'missing masses' are not so missing after all. In discussing the fate of two alternative schools of chemistry in the nineteenth century he states that, 'There is no denying that part of the reason why Liebig [a German innovator in organic chemistry] was a success was because the material world responded with regularity when subject to the treatment given it in his apparatus' (1991: 36). Equality of treatment means searching for a causal explanation of all beliefs, not a commitment that all beliefs are caused by sociology, psychology, biology and so on in the same proportions. As he puts it later on the same page: 'The symmetry resides in the types of causes' – not, that is, in the greatness or smallness of their respective roles. This explanatory role for natural causes is exemplified in a quotation from his discussion of developments in nineteenth-century chemistry. He allows that, 'There is one situation in which it might be permissible to say that the chemistry alone was the cause of a difference, whether in belief, theory, judgement or [other cognitive disposition]. This would be where all the social, psychological, economic and political factors were identical, or only differed in minor or irrelevant ways' (1991: 36–7). In such a case, the sociological variables would have been 'controlled' for. Many sociologists of science who see themselves as proponents of the Strong Programme might object to such a statement. They would argue that in scientific controversies it is impossible to determine which factors are 'minor or irrelevant'. Typically, opposing scientists would depict these factors in conflicting ways; what one side claimed were minor differences would be interpreted as decisive by their opponents, and so on. By contrast, advocates of the progressiveness of science might seize on this quotation and claim that Bloor has offered a characterisation of the scientific method. Science proceeds, they would say, precisely by arranging experiments so that social, psychological and other variables are balanced as far as possible. Even if this is difficult to manage (as it surely is), that doesn't mean one should dispense with this as an idealised objective.

Bloor's case for the Strong Programme gives a resounding endorsement to empirical, purely naturalistic approaches to the study of all the knowledges that societies and communities develop. His work has offered a major antidote to idealistic accounts of mathematical and logical knowledge and his formulation of the impartiality and symmetry criteria has been at the heart of the advancing wave of SSK studies. Ironically, as the Afterword's detailed exposition makes clear (1991: 166), the naturalistic and materialist basis for his antidote means that his sociology of scientific belief is rather less shocking than both critics and believers have usually taken it to be.

A COUNTER-PROGRAMMATIC: THE EMPIRICAL PROGRAMME OF RELATIVISM

Bloor's four tenets functioned as a rallying point for early work in the sociology of scientific knowledge but Collins, one of the few other exponents of SSK at that time, subtly differentiated his position from Bloor's. In two papers he set out a programme with very similar implications but with revealing differences in the way the investigative strategy was justified (Collins 1981a; 1981b). In the second of these he explicitly considers Bloor's tenets and recommends the dropping of the causality and reflexivity demands: causality because it commits the analyst to views about the similarity of scientific and sociological explanations which may be undesirable, and reflexivity because it is important that the sociologists of science 'should go about finding out things about the social world of the scientist in the same sprit as the scientist goes about finding out things about the natural world', something which would be undermined by persistent attention to issues of reflexivity (1981b: 216). More importantly, however, though Collins accepts the remaining two tenets (symmetry and impartiality), he does so largely on practical, methodological grounds. He does not try to demonstrate that anyone seeking to support an alternative view to his own could only find refuge in anti-naturalism. Collins' principal interest is in natural scientific knowledge, not in mathematics and logic; accordingly, he has few expressly anti-naturalistic adversaries. His argument is that, if one studies any scientific controversy or dispute, the facts of the matter are precisely what is up for grabs. Therefore, the truth of beliefs cannot be part of the explanation of the outcome, since the truth is not known by anyone until the outcome has been determined. In particular, the sociologist studying a contemporary controversy has no access to the truth of the matters under dispute; the truth will only be pronounced once the controversy has been resolved. Thus, Collins argues, the analyst of science must avoid 'TRASP' claims, that is claims about the 'truth, rationality, success or progressiveness' of ideas or concepts. Of course, the sociologist of science can report on the participants' own TRASP claims, but

TRASP factors themselves cannot enter into the sociological explanation of the course of the controversy. If scientists involved in a controversy do not know the truth (and they wouldn't be having a controversy if they did), the sociologist plainly has no way of knowing what the truth is either.

With his claims for the viability of a sociology of scientific knowledge anchored more in issues of practicality than in terms of an opposition to anti-naturalism, Collins' programme too assumes a more practical form than Bloor's. According to Collins there are three stages in EPOR, the Empirical Programme of Relativism (see Collins, 1983; Yearley, 1984: 62–7):

1. Revealing the inevitable openness or interpretative flexibility of scientific results.
2. Examining the social processes that are employed to close debates over results.
3. Investigating the connection between these processes and social forces beyond the immediate community of scientists.

The first stage requires that the analyst shows the 'interpretative flexibility' of the scientific data (Collins, 1981a: 6–7). In one sense, this is merely to re-state the fact that a controversy is taking place. If the scientific data, experimental outcomes or results admitted of only one interpretation, there could hardly be a controversy. But Collins wants to argue something a little more precise than this. Collins has, on numerous occasions, made reference to a famous mid-nineteenth-century dispute between two naturalists, Pasteur and Pouchet (see Farley and Geison, 1982). Both men were notable French scientists who were interested in reproduction and the emergence of life. It was well accepted by then that moulds tended to grow on organic media (on cheeses or vegetables for example). The question was, where did these moulds come from: did they grow from spores floating in the air or did these microscopic life forms arise by spontaneous formation? Pasteur (who believed that life arose almost exclusively through reproduction) and Pouchet (who was more favourably disposed to the possibility of widespread spontaneous generation) sought to settle the matter through experimentation. But their controversy dragged on for many years, illustrating the kind of openness or interpretative flexibility to which Collins wishes to draw attention. In this case, we can see that there were several reasons why the evidence remained open. For one thing, the scientists were working at the very limits of their observational and experimental abilities. The air and other materials in the experiments had to be kept free of contaminants which were not visible and which scientists were not entirely sure how to exclude. Thus it was always possible to disregard particular experimental outcomes as the result of unknown contaminants. Second, the scientists were following slightly different experimental strategies and seldom repeated each other's experiments exactly, typically because they sought to do the experiment better than their opponent. Even in the modern world of experimental science, Collins argues, the search for distinctiveness typically drives scientists to vary their experiments slightly in order to try to improve on what has been done before. Thus 'replications' are seldom precise repeats of the original experiment.

One might suppose that these differences would be sharp at the start of a controversy but that they would subsequently diminish. But supporters of EPOR further assert that, as controversies develop, this interpretative flexibility typically does not decline, or at least not to the point where one view is manifestly correct and the other erroneous. There is both a point of principle and a practical issue here. The point of principle is that there can never be strictly logical grounds for forsaking a theory. It might always turn out that some hitherto overlooked factor or some to-date untried experiment would surface and resurrect the theory; this was the weak point in Lakatos' 'sophisticated' theory of falsificationism discussed in Chapter 1. The practical point is that even minor differences in experimental design can keep a controversy going for many years since both sides can typically continue to provide evidence in support of their own view or against their opponents.

In the face of this logical interminability and of the practical obstacles to coming to agreed outcomes, the next query concerns what it is that does account for the practical resolution of scientific disputes. Proponents of EPOR suggest that there is a variety of stratagems which account for closure. In the case of Pasteur and Pouchet, a scientific jury was set up by the Parisian elite of the 'Academie des Sciences' to decide the issue (Farley and Geison, 1982: 21–4). The jury favoured Pasteur, and Pouchet later withdrew from such contests. In practice, an attempt was made to settle this controversy by the pronouncements of the jury members, not by the pronouncements of nature. In other cases the outcome was settled by rhetorical manoeuvres, by the denial of research opportunities to the 'failed' party and so on. Of course, in the great majority of cases these decisions appear reasonable to the people making them. But Collins' point is ultimately that the controversy gets settled because people decide (or are forced) to stop quarrelling or disagreeing, not because nature makes an incontrovertible pronouncement in support of the victorious side.

The third stage in EPOR is not necessarily applicable in all cases. However, Collins suggests that there may be a 'relationship of the constraining mechanisms to the wider [social/cultural] structure' (1981a: 7). In other words, it may be that people decide to stop disagreeing because broader social factors overwhelmingly favour one interpretation; in such a case, conventional macro-sociological factors could be said to be responsible for shaping the outcome of a scientific controversy. But in his initial statement of EPOR, Collins is careful to present this only as a possibility, albeit an intellectually appealing one. He comments that it 'would be very satisfying if the establishment of a piece of knowledge belonging to a modern main-stream science, with substantial institutional autonomy, could be described in terms of all three stages' (1981a: 7). At least in 1981, Collins apparently believed that no such description was yet available. In the main, therefore, the empirical programme of relativism argues that social and cultural factors within the scientific community are the leading explanatory factors in settling the outcome of scientific controversies. In this restricted sense (and it is a *very restricted* sense), it is society not nature which has the final word.

COUNTER-ARGUMENTS

As already mentioned, both Bloor's and Collins' positions (and those associated with authors making related points such as Barnes, 1974) have stimulated a great deal of critical attention. Much of it has been inattentive to the details of their argument. For example, few critics have distinguished between the arguments of Bloor and Collins. The former is arguing for a programme of naturalistic explanation and his principal adversary is non-naturalistic explanation; for him mathematical knowledge and logical insights (knowledge which we seem to possess *a priori*) appear to be the critical cases. Collins is in favour of a methodological relativism, justified primarily by the analyst's inevitable inability to pick the superior case in any actual controversy. His arguments turn on his claims about the logical non-compellingness of any evidence.

One of the more interesting criticisms – because the author clearly grasps what is at stake for SSK – is offered by Newton-Smith (1981: 237–65). He too focuses on the symmetry and impartiality criteria. He accepts that the impartiality criterion holds: all beliefs require explanation. But, he maintains, those explanations are not symmetrical. His argument revolves around a conception of the 'dictates of reason' (1981: 254).

> When what someone would offer as his reason for believing *p* [some proposition] does indeed provide reason for believing that *p*, I will say that he is following the *dictates of reason*. If someone is following the dictates of reason, then showing that this is so ... explains his belief. If he is not following the dictates of reason we shall,

> *ex hypothesi*, have to give a different type of explanation for his believing what he does. Failures to follow the dictates of reason can be divided into those that are rationalizations and those that are not. The latter would include cases of carelessness, lack of intelligence, lack of interest and cases in which the person in question is acting on a hunch and cannot provide any further reason. (1981: 254, original emphasis)

Newton-Smith hopes to supply a way of analysing belief which acknowledges that all beliefs are in need of explanation but which offers a different form of explanation according to whether beliefs are correctly held or not.

To marry naturalistic explanation of all beliefs with a form of asymmetry, Newton-Smith invokes the idea that we have an interest, indeed a 'general standing interest', in following the dictates of reason (1981: 254). The evolution of our species has provided us with a naturally occurring interest in getting knowledge right. Thus, all beliefs are explicable in natural terms but there is an in-built asymmetry between explanations for correct (evolutionarily favoured) and incorrect (evolutionarily disadvantageous) beliefs. Bloor won't have any of this apparently conciliatory talk. Although it makes sense, even within Bloor's framework, to talk of people being careless in their cognitive assessments and so on, he is not tempted by Newton-Smith's attempt to ally naturalism and rationalism. For Bloor, it is not possible to find a path that is part naturalistic, part normative. Newton-Smith's position, Bloor contends, suffers from the afflictions of all Darwinian approaches to philosophical topics, whether in ethics, aesthetics or epistemology: 'such composite positions … are incoherent. They are trying to meet an impossible condition: making reason both a part of nature and also not a part of nature. If they don't put it outside nature, they lose their grip on its privileged and normative character; but if they do, they deny its natural status. They can't have it both ways' (1991: 178). Evolutionary adaptation is aimed at fitness for an environment. Evolution can refine cognitive performance but it cannot explain how the brain can have access to transcendental reality. Bloor sides with anti-naturalists in denying the possibility of this middle way.

But in many ways just as damaging to Newton-Smith's argument is the observation that his guidelines do not assist one in coping with the position in which the analysts of the Pasteur–Pouchet dispute find themselves. Both sides appear to be following the dictates of reason. It is not as though scientific controversies are generally conducted between the painstaking on the one side and the careless on the other (even if that is how insiders often talk; see Mulkay and Gilbert, 1982 and the analysis of scientists' rhetorics in Chapter 6). The tools which Newton-Smith appears to want to wield seem too crude for the job.

PUTTING THE STRONG PROGRAMME AND EPOR TO WORK

Bloor's argument was made to clear a path for the sociology of knowledge into science and mathematics. Bloor himself appears to favour a revised Durkheimian way of exploring that path (see Bloor, 1978; also 1991: 167). His naturalistic agenda, however, has been most forcefully taken up by his erstwhile colleague Barnes and by others who worked with them in Edinburgh. Their approach owes little to Durkheim, more to neo-Marxists, as we shall see in the next chapter. In the meantime, Collins and his immediate colleagues have shown little interest in stage three of EPOR, and have largely confined themselves to documenting Stages One and Two. In recent years Collins and Pinch have published a very widely read collection of edited case studies (1993). This book aims to show in instance after instance that the evidence from the natural world does not determine belief and that consensus comes only when people decide to stop quarrelling about the natural world. They insist that accounts of how the world 'actually is' have no explanatory force in accounting for the outcome of a controversy because those accounts are the outcome of the controversy, not the cause of its resolution. They aim to show how it is persuasion, rather than reason, that closes debate; how successful mobilisation of supporters rather than incontrovertible experimental demonstration quells the opposition; how creating the monopoly over research funding ensures victory in the battle for beliefs.

Yet, seen in another light, one could summarise these two bodies of literature as saying that most judgements about the scientific analysis of the world are made with care and diligence exclusively by members of the scientific community. These scientists may not follow 'the scientific method' but they are not coming to conclusions on a-scientific grounds either. The clear majority of the cases selected by Collins and Pinch for their 1993 collection have no 'Stage Three' element to them; the controversies are settled within the scientific community. Bloor has consistently insisted on the explanatory role of the physical as well as the social world. Even Barnes argued that, 'Everything of naturalistic significance would indicate that there is indeed one world, one reality, "out there", the source of our perceptions if not their total determinant, the cause of our expectations being fulfilled or disappointed' (Barnes, 1977: 25). This point about the comparative moderation of the practice of the Strong Programme and of EPOR is well illustrated by one of Collins' most recent studies.

In the 1990s Collins returned to the study of the gravitational-wave community, physicists who were also the subject of his earliest studies. Gravitational waves are a form of radiation thought to be generated whenever there are gravitational changes. But the radiation is so weak that only the most huge imaginable changes, such as the collapse or collision of stars, would be at all likely to yield gravity waves of sufficient strength to measure with current equipment on Earth. Physicists are confident that such radiation exists; it is predicted by Einsteinian theory in which they have great confidence. But they want to measure it and, eventually, to use it as a way of examining parts of astronomy currently veiled to us. Collins is interested in the sociology of the detection of such radiation. His early work looked at rival gravity-wave-detecting machines. As no one knew how to detect the waves, nor even whether the gravitational radiation truly existed, the experimental community was faced with a conundrum. Any reported positive result could either be a valid detection on a good detector or a false reading on a flawed piece of apparatus. But because neither proposition could be tested independently of the other (it was not possible, for example, to make gravity waves in a laboratory so that detectors could be standardised), the community entered what Collins termed the experimenter's regress (1992: 83ff.). Collins reports how those who claimed early on to detect gravitational radiation were let down and discredited by a series of errors which, though they did not formally disprove the experimental results, led to widespread scepticism, not least because the earliest detectors came to be widely thought of as too insensitive to measure any incoming waves apart from exceptional events close to or in our galaxy. But so strong are the theoretical arguments around gravitational radiation that experimental interest has continued.

Collins' analysis of the experimenter's regress is of repeated importance to the analysis of disputes in the scientific community. It can helpfully be set out as in Figure 2.1 below. An experimenter who accepts the existence of gravity waves will treat positive observations as evidence for her or his theory while she or he will treat negative observations as indicating that the detector equipment was insufficiently sensitive. The gravity-wave sceptic on the other hand will treat negative observations as an endorsement of their position and will discount positive observations as the result of an over-sensitive piece of equipment. For Collins, this is a perfect exemplification of Stage One of EPOR since, whatever the evidence, it cannot compel belief. Any agreement reached in the scientific community will necessarily be the result of a decision to stop disputing and to shelve doubts.

In a study published in 1998 Collins examines the responses of present-day experimentalists working under these unfavourable conditions. They assume that their equipment is unlikely to pick up any waves unless they are extraordinarily lucky and they know that new, more sensitive detectors are to come 'on stream' in the next ten years. Precisely because the detectors are sensitive to all kinds of noise and disturbance, no single detector is likely to be able to make a reliable report of a finding. A large signal

		Believes in gravity waves	
		YES	NO
Observes movement in the detector	YES	*'proves' thesis*	*'evidence' caused by extraneous factors*
	NO	*equipment not sensitive enough*	*'proves' thesis*

Figure 2.1: The 'experimenter's regress'. According to Collins (1992), if there is a novel phenomenon and only one way (or a few ways) to check it, scientists enter this regress.

would be suspect because it is out of line with theoretical expectations. And a signal of the 'right' size could easily be caused by an accidental disturbance to the detector or by gremlins in the equipment. Accordingly, experimenters have responded by working in pairs of teams. If the same 'signal' is observed on two machines separated by thousands of kilometres then that should overcome the problem of accidental disturbances. But of course coincidences will still happen, so an element of judgement is required to sort out the 'real' effects from the spurious ones. Collins reports two further indicative aspects of this case. The first problem is that the waves will not 'strike' the two machines at exactly the same time since, even travelling at the speed of light, the impacts will almost certainly be separated by a tiny fraction of a second. Accurate time-keeping is clearly vital but Collins reports that there have been recurrent problems in the management of time-keeping with some institutions' clocks drifting by a second or more a day and other reports of scientists calling the local radio station to get 'time checks' (1998: 323). It will not be easy to get the natural world to 'speak' of co-ordinated measurements through this fog of uncertain timings.

The second, more sociological complexity is that the two putative partners may have different thresholds for publicising their findings. In other words, some centres are more conservative about when they have a 'real' finding than others. Collins points out that if two centres shared their data freely, the less conservative one can always pre-empt the other by publicising the joint findings as soon as its staff see fit. According to his study, the more conservative scientists have responded by sending their data but with the time/date identifications removed. When the less conservative scientists spot a strong coincidence between the two experiments' printouts they can still publish but they risk humiliating themselves because they cannot be sure that they are comparing the records of simultaneous signals. This hands the initiative back to the more conservative groups even though there is a worry that this removal of time/date labels will cause unnecessary extra work and make true co-incidents harder to spot (1998: 331).

What is particularly interesting about this case study is that, after over 20 years, Collins is still able to find detailed and specific ways in which his original observations about the negotiation of knowledge in this field continue to be borne out. Though these physicists are energetically and tenaciously looking for gravity waves, there are persistent problems in knowing whether they are measuring the phenomenon or not.[2] Early on, the leading question was: whose detector works and whose does not? More recently the issue has become: which of the two linked groups is making the right judgement about which co-incidents are data and which are mere coincidences? In neither case does the evidence from the cosmos (literally in this case) tell them unambiguously. It is in the hands of the scientists, as a group or community, to decide what the facts are, much though they want the facts to be facts about the world. In that narrow and precise sense, the measurement (or not) of gravity waves is the outcome of the community's decision and not the other way round. Basic facts about the cosmos are being decided, for the rest of the scientific community and the broader culture, by a relatively few insiders.

Naturally, members of the gravitational radiation community try to get at other ways of detecting this elusive phenomenon. The experimenter's regress works at its sharp, infuriating best when there is only one technique to measure the one physical process. It may be that one can deduce from theoretical postulates the likely characteristics of the gravity waves from particular kinds of gravitational events. In that case one would not necessarily need two detectors since the 'profile' of the wave would be the thing which sets it apart from accidental spikes. But even then, it would be necessary to make a judgement about whether the deductions from the theory are sufficiently precise, whether the match between the measured wave profile and the anticipated wave form is close enough and so forth. Each of these decisions requires judgement and thus forms only part of the larger regress problem.

Collins' overall conclusion therefore is that, in the end, scientists have to decide on the nature of the physical world. That is not to say that they are free to 'make it up'. Generally speaking, the scientific community sets enormous store by evidence, and scientists' working lives are often dedicated to trying to improve the quality of that evidence. But EPOR suggests that the evidence is never fully compelling. This means that, even within the physics community, there can be cultural differences in scientists' approaches to data. But such differences are likely to be internal to the scientific community and may well not relate to external ideological or material influences. The expert community may be highly insulated from the broader social milieu. To pursue the sociology of scientific knowledge is not necessarily to propose that broad social factors impinge on the shaping of scientific concepts and theories.[3]

EPOR, THE STRONG PROGRAMME AND THE 'REALITY' OF SCIENTIFIC KNOWLEDGE

In concluding this chapter three principal points need to be made. First, though the Strong Programme and EPOR (on the one hand) and philosophical realism (on the other) appear to be directly opposed, they are in a sense intimate enemies. Both acknowledge that the scientific community determines how the natural world is. Realists claim that scientists do this by eliminating social and cultural factors as far as possible – ideally completely. Collins agrees to the extent that he views 'realism' as the natural attitude of the practising scientist; but it is an unattainable ideal. Collins argues that social and cultural factors are ineliminable since, in practice, scientists must decide when the evidence is sufficient. The two positions are mirror-opposites. Collins only has to be a little wrong for realism to be right; realism only has to be a little exaggerated for EPOR to be justified. As will be seen in Chapter 4, this underlying similarity between social constructivism and realism is used by authors such as Latour and Callon as a reason to reject both.

The second point also relates to a claim of Latour, the claim with which this book began about the importance for sociology of understanding society's 'dark matter'. To argue, as Bloor and Collins do, that the compellingness of mathematical proofs and the apparent certainty of scientific knowledge stem from culture (rather than directly from the compellingness of the evidence) leads to a rather unexpected interpretation of the missing masses. They are missing not just because sociology has overlooked them but also because sociologists have not understood how deeply social the 'masses' themselves are. In the second half of this book a key concern will be to see how an appreciation of the social character of the missing masses can aid in the sociological analysis of broader societal issues such as risk and the role of expert advisers on policy matters.

Both Bloor and Collins are engaged in generating naturalistic accounts of scientific and mathematical knowledge. Hence the third concluding point is that their aim is to clarify how scientific beliefs are produced not how they *should be* produced. By contrast, most of the authors reviewed in Chapter 1 were interested in specifying how science ought to be done. But these approaches are not necessarily as distinct as one might suppose. For example, Kuhn's (1977) analysis of cognitive values in the scientific community was primarily a descriptive account: he justified the values by claiming that they were the ones most commonly honoured in the community's practice. Through his commitment to reflexivity Bloor clearly aims to carry out his Strong Programme in a scientific manner. If he were to give advice on how to do that, the obvious place for him to look also is to extant practices within the scientific community. Similarly, Collins' recent emphasis on differing evidential cultures within the gravitational wave community means that his focus is on the various cultures (one might almost say values) within the scientific profession. In this normative arena the descriptive approach to cognitive values and the descriptive ambitions of Bloor and of Collins become rather closer than most commentators appreciate; this issue will be taken up again in Chapter 7.

One final, additional point can be developed from Bloor's Afterword. When challenged to say what the leading 'finding' of the Strong Programme had been, Bloor offered finitism as his answer. The general finding of SSK is that '*all* concept application is contestable and negotiable, and *all* accepted applications have the character of social institutions' (1991: 167, original emphases). Thus, the supposed truths of mathematics and logic do not have their own automatic implications; rather the drawing of implications is always a cultural accomplishment. Similarly, Collins' main claim is not about Stage Three of EPOR (the possible influence of external factors on the development of scientific knowledge) but about Stage Two: the assertion that agreement results from people ceasing to argue, not from evidence compelling people to agree. Finitism in this sense is the key result of the sociological turn in science studies: people collectively determine what knowledge is, even if they experience that knowledge as compelling and external to themselves. This marks a fitting end to this Part; in the next chapters I shall review how sociologists have sought to develop science studies on these foundations. Part II starts with an examination of the most systematic sociological attempt to develop Bloor's fundamental insight.

---

[1]This point is nicely expressed by the sophisticated realist philosopher Hilary Putnam: 'The metaphysical realist in wanting a property that he can ascribe to all and only true sentences, wants a property that corresponds to the assertoric force of a sentence. But this is a very funny property. To avoid identifying this property of truth with that of assertability, the metaphysical realist needs to argue that there is something we are saying when we say of a particular claim that it is true over

and above what we are saying when we simply assert the claim. He wants truth to be something that *goes beyond* the content of the claim and to be that in virtue of which the claim is true. This forces the metaphysical realist to postulate that there is some single thing we are saying (over and above what we are claiming) whenever we make a truth claim, no matter what sort of statement we are discussing, no matter what the circumstances under which the statement is said to be true, and no matter what the pragmatic point of calling it true is said to be' (Putnam, 1994: 501, original emphasis).

[2]The sociological aspects of the detailed work of the physics community have also been explored by Galison (see 1987: 263–78) and by Pickering (1984). As an example, Galison picks up on the consequences for the physics community of 'scale effects': 'physics goals demand an increasing size, but that augmentation creates an increasing delay between proposal and publication that makes it possible that physics goals will change during the course of the experiment' (1987: 265). In exercising judgement, scientists have to bear in mind not only the evidence available but also the prospects for the appearance of new evidence given the lengthy waits for new equipment to come on stream. Such practical exigencies constitute part of what Pickering (1995) refers to as the 'mangle of practice'.

[3]In the gravity-wave case, however, Collins does suggest that some of the experimenters' caution about claiming to detect gravity waves may be due to them trying to avoid jeopardising relations with the teams working on the new, upcoming generation of detectors. If current technology were to turn out to be sufficient to study gravity waves then the new-style detectors would not be so urgently needed. Some physicists may thus have an interest in promoting the idea that conclusive results cannot be generated with the current equipment.

PART II: SCHOOLS OF SCIENCE STUDIES

3 *Knowledge and Social Interests*

INTRODUCTION

The main theory of social interests as applied to the sociology of scientific knowledge was developed by Barnes, MacKenzie, Shapin and fellow authors including Pickering. As they were all based at the Science Studies Unit in Edinburgh, this interpretation of the sociology of science has often been dubbed the 'Edinburgh School'. This title is somewhat misleading, however, both because the majority of these authors are no longer in Edinburgh and because Edinburgh-based Bloor was never a practitioner of this approach in the strict sense. Furthermore, as we shall see, no single theory of social interests has been successfully stabilised. Various 'Edinburgh' authors use the term in differing ways and there is no single exemplary study which serves to represent the mainstream position. Despite this, the idea of explaining the development of scientific knowledge through the operation of social interests marked the first comprehensive theoretical position in the sociology of scientific knowledge. Furthermore, it also effectively marked the first systematic attempt to carry out something resembling Collins' 'Stage Three' of EPOR. This chapter will begin with the background to interests and then move on to a clarification of the Edinburgh conceptualisation of knowledge interests. A case study example will follow, then critical commentary leading to a concluding assessment.

LINKING BELIEFS AND INTERESTS

The connection between beliefs and interests is easy to understand, at least on the face of it. In daily life, people often seem to believe things which suit them. Wealthy members of society commonly believe that income tax rates should be low, not – they say – because it would make them even better off, but because it's the right policy for society. It would stimulate the most entrepreneurial and commercially talented people to work even harder, thus expanding the economy and benefiting everyone in the long run. They take it that their belief in low taxation is well supported by economic arguments and by evidence about successful economies elsewhere in the world. Similarly, labour groups or trades unions often favour protectionist measures not – they say – merely because it will safeguard their members' jobs, but because the evidence is that, as a matter of fact, other countries which engage in stealthy forms of protectionism prosper. They argue that engaging in truly free trade is often a mild form of national financial suicide. Supporters of fox-hunting in Britain argue, and seem to believe, that their sport is actually beneficial to foxes since, without the local motivation to maintain a healthy fox population precisely in order to have animals to hunt, farmers would be inclined to shoot foxes and to cultivate the unfarmed pieces of land (the 'cover') where foxes can conceal themselves. Hunters present themselves as the fox species' true friends. Of course, one may doubt the sincerity of such beliefs; they may just be rationalisations designed to cover up for self-interest. But in some cases at least, it does appear that people within a relatively unified social group come to develop sincerely held, if not always very critically examined, beliefs which support and legitimate their interests. Without, as yet, being very precise about the terms we use, we can say that beliefs and interests are linked in a mutually supporting way: beliefs legitimate interests while interests support the sub-cultures within which the beliefs flourish. This relationship is represented in an approximate way in Figure 3.1.

Naturally enough, it is uncommon for people to talk about their own beliefs in this way. While members of the hunt deploy their belief in the environmental benefits of fox-hunting to defend their interests, they do not suggest that their beliefs are sponsored by those interests; they argue that their beliefs are justified by the facts or by their personal experience or by the testimony of country folk. It is their political opponents who draw attention to the interests-to-belief connection. This issue is bound up with Bloor's notion of symmetry, as will be discussed later.

So far I have dealt only with anecdotal examples and with items of belief which could be interpreted as 'matters of opinion', that is issues about which varying views can reasonably be held. What Barnes and his fellow authors did was to argue for the extension of this mode of thinking to detailed case studies and to issues of scientific belief where one would expect there to be intense efforts to generate decisive evidence and where, therefore, one would anticipate only one correct answer.

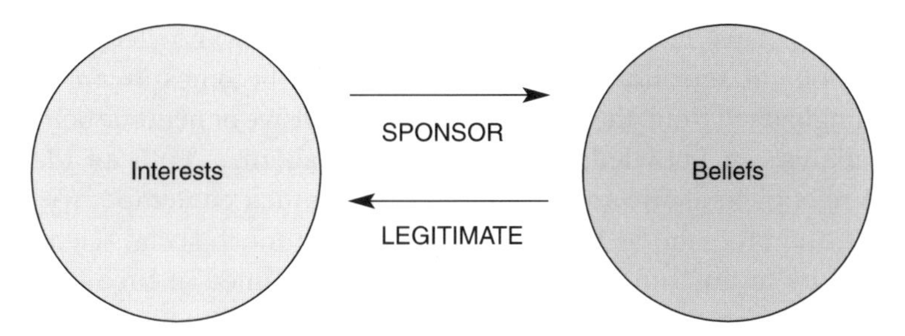

Figure 3.1: The mutually supportive relationship of knowledge and interests.

Barnes and colleagues were not the first sociologists to try to develop a theory of the links between interests and beliefs as Barnes happily acknowledges (1977: 11). Marxists such as Lukács had suggested that members of social classes had interests in common and thus would typically hold shared beliefs which meshed with their class interests (see Lukács, 1971). But it was the critical theorist Habermas who did most to advance the idea of interests as a central analytical concept for social studies of science. Habermas' aim was to clarify the basis for a 'critical theory of society', that is, a theory which is at the same time empirically robust *and* progressive in its political and ethical implications. The model for such an enterprise was Marx's political economy of capitalism which claimed both to describe how the capitalist economy worked and to indicate how it could be superseded. But Habermas acknowledged that Marx and subsequent writers had not adequately clarified what it might mean for a theory to be both empirically accurate and liberatory; in any case, straightforward Marxism seemed rather defective in the accuracy department.

In Habermas' view, his search for a solution demanded a review of the basis of the various kinds of knowledge. In outline, he argued that all human knowledge was developed in relation to species-wide interests (1972). In that sense, we do not have knowledge for its own sake, but knowledge that serves enduring interests. Habermas argued that scientific and technical knowledge was fundamentally related to an interest in prediction and control of the natural world. Other forms of systematic knowledge – literary and art criticism for example – were not oriented to such an aim. The point of literary criticism was not to predict what Salman Rushdie would write next, but to enhance understanding of the meaning of creative art-works. Having established that systematic knowledge was not subservient to one interest only, Habermas then introduced a third interest, an emancipatory interest, which is supposedly served by the social sciences. On this view, the appropriate goal of the social sciences is not to predict and control behaviour but to identify the factors which constrain people's conduct, allowing them to reflect on and overcome these constraints. Habermas' model for this kind of knowledge was the psychoanalytic encounter in which the patient becomes aware of, and accordingly freed from, the background to compulsive or neurotic behaviour. This analysis of knowledge interests gave Habermas both an ideal for social science to aspire to and a criticism of much contemporary social science; the problem, as he saw it, was that too much social science was directed by an instrumental interest, implicitly aimed at predicting and controlling people's conduct.

Despite his initial enthusiasm for such a model of a critical theory, Habermas later departed from this line of thinking (1973).[1] He came to doubt whether reflection was a suitable model for political emancipation; emancipation might require more action than the ideal of reflection implied. In addition, there appeared to be certain constraints – the constraint of linguistic structures for example – from which it made no sense to try to liberate ourselves. Linguistics, for example, is thus a social science which aims at reconstructing the basis of a set of human skills but it does not aim at having us transcend language; it is not emancipatory in the way that Habermas originally envisaged. Habermas forsook his three-fold theory of epistemological interests. But, for present purposes, the attractive feature of his work was that Habermas appeared to have come up with a notion of interests which was compatible with objectivity. According to his theory, science and technology were founded on a human interest but were also, within the appropriate domain, capable of objectivity. Indeed, the interest is actually the precondition of objectivity. Knowledge would not be better without the interest; it would be aimless. And it was precisely this point that created the opportunity which 'Edinburgh' interest theorists sought to exploit.

SOCIAL INTERESTS AND SCIENTIFIC KNOWLEDGE

The 'Edinburgh' interest theorists adopt a variant of Habermas' argument. For Habermas, our interests in prediction and control and so forth are species-wide; he refers to them as quasi-transcendental (recalling Newton-Smith's argument about our 'standing interest' in control over the natural world discussed in Chapter 2). But, even for Habermas, there is some possibility of contingent, empirical variation in the way these interests play out. To have his theory work as critique at all, he has to put forward the idea that we may confuse these interests, notably by applying the instrumental interest to the social sciences. Still, he is not at all specific about why this has occurred or about how it is possible to 'mix up' allegedly transcendental interests. Habermas thus allows a small (though far from fully specified) amount of empirical variation in people's responses to these interests. By contrast, 'Edinburgh' interest theorists wanted to develop a systematic and fundamentally empirical research programme based on this insight. If you are going to examine the reasons for a scientific controversy or wish to account for the emergence of 'sides' or competing schools of thought, then a common interest will not be much help. Interest theorists have instead sought to identify conflicting interests underlying the opposing sides in scientific disputes. The guiding idea is that scientists (and others) generate claims to knowledge in the light of their interests and that controversies arise when interests conflict. The nature of the argument here can best be understood by looking at a justly renowned case study.

MacKenzie (1978, 1981) examined a controversy which took place in Britain in the early years of the twentieth century over the best method for assessing the statistical connectedness of nominal variables. Nominal variables, such as eye colour or choice of daily newspaper, are distinguished from other variables analysed by statisticians because they cannot be assigned a precise numerical value (as can be done with, say, weight) nor, in many cases, even be ranked (as might be done with political-party allegiance on a left–right spectrum). Two figures were prominent in the dispute; each developed his own statistic and argued for its superiority over that proposed by his opponent. In explaining how the controversy arose and developed, MacKenzie suggests that the two statistics were related to the practical agenda of the disputants and were thus sponsored by specific, empirical interests. The more well-known figure was Pearson. He had already conducted pioneering work on establishing the correlation between measurable variables such as height (known as interval variables). He now introduced a statistic, the tetrachoric coefficient (referred to by MacKenzie and hereafter as r_T), which expressed the correlation between two nominal variables or – to be more precise – between two variables each of which had two discrete categories (let's say parents and children with brown or blue eyes). This statistic was relatively complicated, but Pearson claimed that r_T was the best way to measure the association between these variables because it was as compatible with existing statistics as it was possible for such a measure to be. Moreover, he defended r_T because all the values which the statistic could adopt were meaningful. It could have any value from +1 to –1 and all those values ought to have a precise meaning. For example if r_T came out at +0.6 then that should mean the correlation was twice as strong as the case in which the value proved to be only +0.3.

His opponent, Yule, devised an alternative statistic, Q, which was easy to calculate. An example will clarify how Q functioned. Imagine that some people in a town are vaccinated and others are not; subsequently some people contract the anticipated disease and die while others are exposed to the illness but live. Yule's Q is set up so that it will equal one if the vaccine is highly successful, meaning either that all people who received it live (while some others died) or that all who did not receive it died (while some of the inoculated lived). If the vaccine makes no difference to one's chance of survival, the statistic equals zero. The statistic will record –1 in the unhappy event that all the vaccinated people die while some of the unvaccinated live, or that all those who escaped the vaccine live while some of the treated die. If, as is likely, the inoculation shows some positive results but without being so effective that, say, only the vaccinated have a chance of living, then Q will have a positive value between 0 and 1.

Both leading figures, and their associates, understood how to use all the statistics on offer but according to MacKenzie they persisted in seeing their worth very differently. Yule objected to Pearson's statistic because it assumed that beneath the two-by-two tables there lay a pair of normally distributed variables. For example, one could produce such a table by categorising fathers and sons (at the time statistical work tended to privilege males) as short and tall. One could then ask how strongly being a tall father was associated with having tall sons. But, of course, people in the tall category would actually represent a range of heights, so the categories would be 'slices' out of an underlying statistical distribution. In a world without measuring tapes this might make good sense, but Yule claimed that nominal variables were often not like this. To use the earlier example, Yule insisted that people are either vaccinated or not, and

> all those who have died of small-pox are all equally dead: no one of them is more dead or less dead than another, and the dead are quite distinct from the survivors. (cited in MacKenzie, 1981: 162)

Furthermore, even where it might make sense to assume that the categories related to some underlying variable, there was no general reason to suppose that the underlying variable was normally distributed. Finally, Yule attacked r_T using re-calculations from some of the tables offered by Pearson. Returning to our tall and short male relatives, it ought not to matter to Pearson's statistic precisely where one drew the dividing line between tall and short, whether at 173 cm (5′ 8″) or 178 cm (5′ 10″). The strength of the association, and thus the value or r_T, should be the same. Yule argued that the value did change and that, therefore, the much vaunted consistency of r_T was illusory.

According to MacKenzie, Pearson was equally critical of Yule's approach. Apart from the values that were stipulated in advance (+1, 0 and –1), none of the values which Yule's statistic generated had any scientific meaning. Two values of Q could not be compared on any straightforward mathematical basis. Worse still, though the values of Q looked like values of the correlation coefficient (r) used for correlating interval variables, they bore no systematic relationship to each other whereas, of course, r_T was directly comparable to r since that was how Pearson had set it up in the first place. Finally, Pearson and his colleague Heron noted that many categories were aggregates of underlying variables: some sick patients were sicker than others, some people injured at work suffer worse wounds than others. Accordingly:

> [they] justified the biometric [i.e. their school's] position by arguing that it was necessary to have *some* hypothesis about the nature of the continuous frequency distribution of which the observed classes were groupings. In practice, they argued, methods based on the normal distribution almost always gave adequate results. (MacKenzie, 1981: 164)

Thus, though they understood each other's work and were adept at using their rival's statistics, these leading statisticians viewed the world differently. As MacKenzie insists, 'both sides felt that the theory of the other was *wrong*, and not merely *misapplied*' (1978: 60, original emphases). Commenting later in a paper written with Barnes (Barnes and MacKenzie, 1979: 58), he notes that the protagonists 'were, if one wishes to put it so, arguing as if they lived in different worlds'.

MacKenzie then seeks to account for these competing statistical visions. Since both men were accomplished statisticians and since both statistics were in some sense successful, one cannot reasonably account for this case in terms of one protagonist being correct and the other mistaken. Accordingly, the analyst must look outside of statistical reasoning and evidence for the explanation for these differences. MacKenzie suggests that it makes good sense to account for the differences in terms of participants' interests since each figure produced a statistic that was in conformity with his interests. The case for Pearson is the easier to comprehend: his 'work in statistical theory continued the link … between the mathematics of … correlation, and the eugenic problem of the hereditary relationship of successive generations' (Mackenzie, 1981: 168). The principal obstacle he faced was that many biological features, and particularly the 'eugenically crucial mental characteristics' (1981: 169), could not be quantified. Before standardised measures of 'intelligence' were generally available, Pearson could only get data about mental brightness, conscientiousness and so on in terms of broad categories. For example, teachers could be asked to classify children according to their perceived intelligence.[2] Once Pearson had these data he could calculate the strength of the correlation between relatives' intelligence using r_T. Then, by comparing r_T with existing figures for the correlation (measured as 'r') between the same types of relatives' heights, he could hope to establish whether intelligence and other mental characteristics were as strongly inherited as other biological features. The close relationship between r and r_T was plainly critical to this enterprise. Furthermore, according to MacKenzie, the advancement of the eugenics movements was one of Pearson's broader social interests, which related in turn to his membership of the rising professional classes. These classes were anxious to separate themselves from the mass of working people and to gain influence by establishing their central importance to a successful, competitive society. The programme of eugenics, which advocated selective breeding in order to raise the level of (supposedly inherited) intellectual abilities, depended on the esoteric knowledge of professional people and was readily attractive to men such as Pearson. Thus, MacKenzie presents Pearson's class position as fostering in him certain political views which led to his support for the 'scientifically based' political programme of eugenics. As a result of this, he was led to see the need for statistical measures which were consistent with his opinions on the distribution of intelligence in human populations. Moreover, MacKenzie cites evidence that Pearson's political commitments preceded his derivation of the tetrachoric coefficient. In terms of Figure 3.1 (above), the interest sponsored Pearson's belief about the right way to measure association; in turn, this measure legitimated his socio-political interests by enabling him to claim authoritatively that intelligence was highly heritable and that selective breeding would raise societal levels of intellectual ability.

MacKenzie acknowledges that the link between Yule's interests and beliefs is less overt. Yule appeared to be antipathetic to eugenics and thus had no particular interest in unifying the treatment of nominal and interval data. His principal practical concerns in the statistical field related to issues such as charting the amount of destitution and miserable poverty in the population, where elaborate statistical theory was not needed but where readily calculable measures which could be used in a pragmatic fashion, case by case, were of value. The measures he viewed as valuable, such as Q, meshed with these practical objectives. Finally, MacKenzie proposes that Yule's statistical preoccupations, in favour of ameliorative social interventions to help the destitute and against eugenic planning, can readily be tied to the interests and outlook of downwardly mobile, patrician conservatives. Such figures were at odds with the ambitions of the professional classes; their idea of reform was targeted at removing the greatest stimuli to unrest among the lowest stratum in society. Evidence about Yule's family background and career is employed to locate him in this formerly-elite, downwardly mobile sector; thus his social interests appear to match the cognitive interests which his coefficients seem best to serve. MacKenzie summarises his argument as follows:

> the two divergent approaches to the measurement of association to be found in the work of Pearson and Yule can be seen as expressing different cognitive interests; that these different cognitive interests

> arose from the different problem situations of a statistician whose primary commitment was to a research programme in eugenics and a statistician who lacked any such strong specific commitment; and finally, that eugenics itself embodied the social interests of a specific sector of British society, and not those of other sectors. Thus differing social interests can be seen as entering indirectly, through the 'mediation' of eugenics, into the development of statistical theory in Britain. (1978: 71)

EXPLAINING BY USING SOCIAL INTERESTS

MacKenzie's case study has been the subject of a great deal of discussion, initiated both by critics (Woolgar, 1981; Yearley, 1982) and also by the supporters of the interest approach (Barnes and MacKenzie, 1979; MacKenzie and Barnes, 1979). In this section, three main arguments about the value of this case as supporting evidence for the general theoretical position will be reviewed. Before that, it should be noted that using the Pearson–Yule story in this way is not necessarily to endorse all details of the case study. For example, a re-consideration of the evidence might give grounds for thinking that the protagonists could be seen as being more accommodating to each other than it appears in MacKenzie's account (see Yearley, 1982: 368). However, this is an aside and my argument does not depend on this kind of re-evaluation; the following critical points presuppose that the details of the case study are fundamentally agreed.

The first difficulty which this case presents for the theory relates to the nature of the connection between knowledge and interests. In his case study MacKenzie is generally cautious in the claims he makes. He is cautious both about his identification of cognitive interests, which he describes as 'tentative' (1978: 48 and 66), and about how much (and in what precise ways) interests drive the production of belief. MacKenzie explains that 'Pearson's approach ... was evidently structured by an interest' in unifying the treatment of nominal and interval data (1978: 49). But in this sentence it is unclear what 'structured' really means. 'Structured' appears to straddle two options: either interests impel actors towards the adoption of certain approaches and strategies or actors actively interpret their interests in the course of arriving at their conclusions. The former position is materialist and appears to be in line with the starkest demands of the Strong Programme, while the latter avoids the hazards of determinism, thereby maintaining a degree of autonomy for cognitive processes. It seems clear that, in the course of his case study, MacKenzie favours the latter interpretation. The contestants in his case study engage in revisions of their statistics and they respond creatively to each other's criticisms and counter-arguments. However, two related problems have been detected in the line of interpretation apparently favoured by MacKenzie; both concern the connection between interests and the knowledge they sponsor. Brown (1989: 55) makes the point that, once it is admitted that the connection between knowledge and interests is interpretative, there is no inevitable connection between interests and the beliefs they sponsor. Support for a belief is not so much deduced or otherwise automatically derived from an interest but is an interpretative accomplishment. The interest theorist's response to this is, reasonably enough, that the connection does not have to be – indeed cannot be – an automatic one. All that is required is that, in each case, the two sides happened to have figured out that specific beliefs assist them in advancing their agendas (see Barnes et al., 1996: 121). Yule did not have to favour Q but (interest analysts want to say) the historical finding that he did advocate it can be accounted for in terms of the way it advanced his interests.

But this leads to the second, closely related point: that for the knowledge and the interest to be linked there has to be some sort of 'fit' between them. The actors have to agree that theory X is in the interests of group Z. And, to grasp this fit, one is committed to a view that actors can somehow rationally appraise the implications of their beliefs or, to put it another way, that there can be a 'right' answer to the question of whose interest is served by which belief. If there can be this level of intellectual agreement, a fact which the practice of the Edinburgh-School explanation takes for granted, then why can't the rest of actors' beliefs be explained in conceptual terms also? Bloor's response to this kind of criticism is essentially to banish it:

> Interests *don't* have to work by our reflecting on them, choosing them, or interpreting them. Some of them, some of the time, just *cause* us to think and act in certain ways. The real basis of the objections to interest explanations is the fear of causal categories. It is the desire to celebrate freedom and indeterminacy, and the reluctance to construct explanations rather than simply describe. (1991: 173, original emphases)

The difficulty is precisely that, whatever ability people may have to know what their interests are, surely they have the same ability to know things about the natural world.

In Edinburgh-School explanations, the evidence available to participants is first shown to be ambiguous, just as in Stage One of EPOR (discussed in the last chapter). In the case chosen by MacKenzie, there is more than one way of dealing with troublesome nominal variables. But then it turns out that the actors, though unable to agree about which is the better statistical approach, are both able to grasp unambiguously which statistic better promotes their interests. Making actors the interpreters of interests in this way also means that the analyst is unable readily to accommodate the possibility that the perspective itself could become contrary to the actors' interests. It seems to wed Yule to Q wherever the adoption of Q might lead, and so on.

This feeds directly into the second main critical point. As well as some lack of clarity about how precisely knowledge and interests relate, in Edinburgh-School explanations there is a lack of precision about the meaning of instrumentality, a term which – as noted above – Barnes and other theorists took over from Habermas. Again the difficulty is to do with the issue of interpretation. Put at its simplest, the question is whether:

- there are simply contingent, historical interests; OR
- there is an instrumental interest alongside such other assorted interests; OR
- there are many instrumental interests.

From an interest-theory point of view, one could easily imagine a study along the lines of the first of these possibilities. Religious or superstitious beliefs, which are not amenable to straightforward testing, might be thought to conform to the first of these possibilities. Within the Christian tradition there have been long and heated arguments about, for example, the nature of the 'Trinity': what exactly does it mean for God to be three-fold. At times, different interpretations have appealed to, and been taken up by, different factions and there have been bloody confrontations between these factions which have been put down to disputes over the interpretation of the Trinity. Some of the contestants, at least, presumably held sincere beliefs about their own rightness and their opponents' error. That one's opponents are promoting falsehoods about God might well be seen as a reason for wanting to punish them, even with violence (even though religious ideologies can also provide excellent cover for politically motivated aggression). But one can suggest that it is the persistent inability to discover the answer to this question, at least in this life, which allows the debate to persist; the sides can argue with each other interminably precisely because the debate cannot be conclusively settled. Barnes and colleagues have, by and large, not wanted to make this kind of argument about scientific and mathematical knowledge since they want to do justice to the sense of objectivity which science elicits and to the commitment to empirical testing which the scientific community characteristically displays.[3]

On other occasions Barnes has adopted a position much closer to Habermas' original formulation. On such occasions Barnes' argument is that actors have many interests in making claims about nature, one of which is an instrumental interest. Scientists may commonly deny that they have any interests in this sense at all; they may claim to be disinterested. But he asserts that:

> the 'disinterested evaluation' of knowledge is in most contexts a harmless enough formulation, which can be taken as practically equivalent to 'evaluation in terms of an authentic interest in prediction and control'. (1977: 91)

The idea that one can identify an 'authentic' interest is at odds with a later view offered by Barnes and MacKenzie (1979: 52) according to which scientists may 'differ ... in the instrumental interests which pre-structure their evaluations'. Indeed, this appears to be the position adopted in MacKenzie's case study. Pearson and Yule had different objectives; the statistics they derived and the tests that they employed to gauge the value of those statistics were closely tied to those objectives. But the difficulty here lies in understanding what it means for these interests to be different and yet both instrumental. Adopting this two-fold interpretation pays dividends for interest theorists because they are able to assert that scientists are concerned with grasping the world, yet the same scientists' beliefs can also be ideological because there can be variations in actors' understanding of instrumentality. But, without greater attention to how discrepant versions of instrumentality can arise and persist, this compromise seems uneasy. Again, Bloor's response is to side-step the issue, arguing that 'Undeniably the terminology of interest explanations is intuitive, and much about them awaits clarification, but instead of seeing these as practical difficulties their critics see them as weaknesses of principle' (1991: 170–1). But the continuing lack of clarity around this point, which has persisted for well over a decade, is perhaps indicative that the difficulties are not merely practical but rather more deep-rooted.

The final conceptual difficulty with interest-theory relates to the extent to which interests allow us to understand the outcome of a controversy. MacKenzie presents evidence to suggest that Pearson and Yule developed opposing statistical approaches (structured by their interests) and that neither was persuaded by the other's arguments. Even if we accept this much, there still remains the question of what happens to scientific belief in the long run. A conventional (even a Popperian) understanding of science could allow that rival objectives would give rise to contending beliefs. But the rationalist (such as Newton-Smith) would take refuge in the assumption that scientific testing would gradually eliminate the influence of those interests or at least whittle beliefs down to the ones which most closely accord with our interest in prediction and control. For this reason, an ideal case study for an interest theorist would be one where the participants' interests overwhelmingly shaped the outcome of the controversy. In this instance however, the controversy was never settled in a definitive manner (MacKenzie, 1981: 179). Scientific attention drifted away from the underlying issue so that neither belief came to be accepted as exclusively or overwhelmingly correct.

In this case there is a further complicating factor. As MacKenzie notes, 'Contemporary statistical opinion tends to deny that any one coefficient has unique validity' (1981: 179–80). He associates this view to some extent with Yule's own outlook, though to do so is to play down MacKenzie's earlier insistence on the extent to which Yule viewed Pearson's statistic as simply wrong. The more compelling point is that this case is anomalous precisely to the degree that it describes a situation where rival beliefs can both, in some sense, be regarded as right. Since both parties to the dispute produced a number of coefficients with differing values, it seems reasonable to conclude that they accepted this pragmatic aspect of mathematical measures to some extent also. Of course, it might be argued that this is an unfair point to raise because I chose to focus on MacKenzie's study in the first place (see MacKenzie, 1984). However, Barnes and MacKenzie themselves (1979: 54–5) present this case as an exemplary one. Accordingly, if knowledge and interest are not closely, deductively linked and if interests do not determine the ultimate evaluation of the knowledge claims, then the explanatory force of interest theory seems rather attenuated. It would seem from this case there is nothing to stop the next generation of statisticians simply getting on and developing statistical thinking with no lasting influence from the interests that apparently informed the Pearson–Yule debate. Of course, this view would not be unwelcome to Newton-Smith, but it hardly suits the original purposes of Barnes nor of Collins or Bloor.

THE STATUS OF INTEREST THEORY

Interest theory was of critical importance to the SSK project because it marked the first (and best) attempt to develop a theoretical vocabulary tied to practical case studies in the sociology of scientific knowledge. Where Bloor and Collins (and Barnes, 1974) had sought to set out the reasons why SSK should in principle be possible, and where Collins had developed case studies illustrating EPOR Stages One and Two, interest theory aimed to introduce theoretical notions which could be employed across numerous case studies. They could be used in the analysis of Collins' Stage Two, by examining conflicts between scientists as indicative of conflicting disciplinary interests or competing professional interests (Dean, 1979; Pickering, 1980; 1984). And they could be employed in Stage Three studies, taking social and political controversies right into the heart of scientific disputes, by suggesting – for instance – how competing class interests supported and promoted alternative cognitive interests and thus different scientific beliefs (see also Shapin, 1979). However, the explicitly theoretical orientation of the Edinburgh School also turned out to be a weakness. Problems with the identification and conceptualisation of interests dogged the programme. In empirical case studies, the difficulty of disentangling long- and short-term interests became apparent. And the precise connection between knowledge and interests was never specified to the extent that it satisfied the whole SSK community. Such conceptual challenges to interest theory were critical since, if interests were going to be the basis for widespread explanation in SSK, they had to be robustly defensible in theoretical terms. Without firm theoretical arguments it was always possible to treat the case study evidence as anomalous or exceptional in some way and thus as no proof that interests were of *general* explanatory importance. Finally, the difficulty of demonstrating that interests were responsible for the outcome of scientific controversies was also widespread.

These difficulties notwithstanding, the attraction of interest theory is easy to appreciate. Explaining the appeal of knowledge-claims in terms of people's interests makes intuitive sense and the theory offers SSK the tantalising prospect of analysing how science is able to appear simultaneously instrumental and ideological, both objective and partisan. The next approach to be reviewed starts out from discontents with the model of interests implicit in this theory.

---

[1]This does not imply that Habermas changed his mind after only one year; his book on interests had been published in Germany in 1968; it was published in translation only in 1972.

[2]It was, understandably, difficult to get data on sons and fathers since fathers had already left school, so Pearson had to study twins and siblings (MacKenzie, 1981: 171). It should also be noted, as MacKenzie observes, that in considering such data neither Pearson nor Yule made a systematic distinction between what we would today think of as the statistics of a sample and that of a population, even though arguments about sample biases did enter their disputes.

[3]Such interest-based beliefs were famously satirised by Jonathan Swift in his tales of Gulliver's travels. In the minute land of Lilliput, Gulliver finds that a political debate rages between those who believe that boiled eggs should be eaten from the pointed end (the establishment view) and the rebels who favour the flat end. Much unrest has resulted from this quarrel resulting in the death of at least one emperor. In this work Swift is lampooning not debates about the Trinity but quarrels between Protestants and Catholics over the transubstantiation of the 'host' during communion.

4 Actor-Networks in Science

ACTOR-NETWORKS AND ENROLMENT

In important ways Actor-Network Theory (ANT) resists summary. It did not set out from a fundamental and unchanging programmatic statement in the way that the Strong Programme or EPOR did. Moreover, Latour's leading methodological injunction is to 'follow scientists around' (1987: 97), which sounds attractively simple but is also beguilingly vague. Worse still, ANT is a conspicuously moving target. The two authors principally responsible for this approach, Latour and Callon, have followed by no means identical intellectual trajectories and they have responded to some critics by insisting that the work addressed in the critiques was not representative (Callon and Latour, 1992: 344; this was a response to Collins and Yearley, 1992a). As one of those criticised critics (see Collins and Yearley 1992b) I have striven in this chapter to be attentive to my reasons for choosing particular examples and have carefully followed up leads given in recent publications by Callon and by Latour regarding the studies which they appear to treat as canonical or representative (see Callon, 1995; Callon and Latour, 1992; Callon and Law, 1997; as well as Latour, 1999b and 2000). There are, though, two convenient starting points for understanding this body of work: one is some methodological observations made by Callon at the start of a widely read case study (1986, reprinted in 1999) and the other is an early article, critical of interest theory, co-authored by Callon.

To begin with the latter, Callon and Law (1982) observe that interest theory is in some senses one-sided and static: static in that it treats interests as relatively stable attributes (either internal to actors or resulting from actors' circumstances) and one-sided in that it traces how interests affect cognition but not the reverse. To remedy this deficit these authors propose that a more dynamic notion of interests should be used. On this interpretation, actors' interests are themselves the outcome of negotiations and interactions. People become persuaded of interests they may 'have', at least in part in light of the kinds of knowledge claim on offer to them. To capture this more active sense of eliciting interest they use the term enrolment: enrolment is the result of the activity of interesting someone in something. With this more flexible term they are able to propose that interests drive knowledge but also that the circumstances of knowledge production may iteratively shape interests: 'we are concerned with the manipulation and transformation of interests, since we see all social interests as temporarily stabilized outcomes of previous processes of enrolment' (1982: 622). Callon and Law also introduce the term 'translation'. An actor may enrol another if she (the first actor) proposes that her knowledge is a means for the second actor to achieve his objectives. The advancement of the first actor's ideas or theory is now seen as being in the second actor's interests. His (the second actor's) interests have now been translated into hers and his original interests have been subtly reconfigured. His interests have been preserved but also modified. Writing about Pasteur's mid-nineteenth-century attempts to enrol farmers in his work on infective micro-organisms, Latour makes the same point. Pasteur offered the farmers a way of reducing animal disease, but only if they accepted his advice and his ideas. The advancement of his intellectual work becomes in their interest; in this sense the farmers' 'interests are a consequence and not a cause of Pasteur's efforts to translate what they want or what he makes them want' (1983: 144; see also 1988a). For Callon, therefore, the sociology of scientific knowledge is fundamentally a sociology of translation. Interests are as much the outcome as the origin of these translations. This revised conception of interests then feeds into Callon's re-working of the methodological basis of science studies.

In his well-known case study paper about marine aquaculture, Callon (1986) criticises writers such as Bloor and Collins for failing to fulfil their own demands for symmetry. He points out that they leave one final asymmetry intact, that between the social (or sociological) and the natural. For Collins to be able to say that scientific findings are always open and that the closure of a scientific debate comes through processes of social negotiation, he has to be able to make a clear distinction between the social world, which is capable of being decisive, and the natural world, whose voice (so to speak) is always subject to interpretative flexibility. The same is true of interest explanations: the social component (the interests) decides the interpretation while the evidence from the natural world is refracted though the actors' knowledge-interests. Callon and Law's idea that interests are, in part, the consequence of translation and enrolments already threatens the primacy of the social over the cognitive. But this argument can be taken further. The very distinction between the natural and the social, between Stages One and Two of EPOR, can itself be treated as a construction. Surely, argues Callon, a thorough-going symmetry requires humans and non-humans to be viewed symmetrically and a symmetrical analysis would require the distinction itself to be seen as a construction: 'given the principle of generalized symmetry, the rule which we must respect is not to change registers when we move from the technical to the social aspects of the problem studied' (Callon, 1986: 200; this line of thinking also underlies Latour, 1993). Callon and Latour's programme has proven attractive partly because of its emphasis on the dynamic interplay of enrolment and the 'translation of interests', partly because of the claim to complete the attainment of symmetry and partly because it appears to re-introduce the natural world, or at least to bring the social and natural worlds back jointly into the sights of the analyst. Their programme seems to shepherd the missing masses back into the social study of science.

GENERALISED SYMMETRY IN ACTION

Callon's best-known case study deals with attempts to farm scallops off the north coast of Brittany in St Brieuc Bay.[1] According to Callon, this fishery had only been systematically exploited since the 1960s, but numbers of the bivalves were already in decline at the time of his investigation. Scallops are fished also in Normandy (north and east of St Brieuc) and in the far west of Brittany, near Brest. The scallops had suffered worse in the Brest fishery since the local variety is fished year round. French consumers are thought to prefer scallops to be 'coralled', that is to contain both the white meat, which they always have, and the orange reproductive element known as the coral. The Brest scallops are constantly coralled whereas those in St Brieuc Bay lose their coral in spring and summer and were thus left alone for several months of the year. The decline in scallops was still a cause for concern in St Brieuc, especially as the prevailing view was that scallop 'farming' – as opposed to harvesting a natural fishery – was not practicable. Rather little was known about scallops but, unlike mussels and oysters, it appeared that they could not be 'reared' and successfully harvested.

However, as Callon reports, some experimental trials in Japan suggested that, with the right sort of protection and tethering, it was possible to encourage scallops to grow and to hang around long enough to be harvested. A team of French fisheries scientists from Brest arrived in St Brieuc Bay with a plan to raise scallops. Although this is not a 'controversy' in quite the sense reviewed in earlier chapters (over ways to measure statistical association for example), there are some common elements. There are two views in conflict and people need to decide which view is correct: either scallops will anchor themselves in the hatcheries according to the fishery scientists' new plans or they will not. Callon suggests that the approach he adopts to this case could be applied in other cases of controversy and dispute also.

Callon claims that the venture went through a four-stage process; he describes these stages as the four 'moments of translation' (1986: 203). The first 'moment' is the process of problematisation. Initially, the proponents of the new strategy have to propose ways in which it is in other groups' interest to align themselves with the new undertaking. If scallop-fishers want to safeguard their economic future, they must overcome the problem of dwindling stocks and this they can do by associating themselves with the new, experimental hatcheries programme. In the language introduced earlier, the desire among the fishers to sustain their fishery is translated into acceptance of the scallop-rearing venture. Similarly, the survival prospects for the scallops themselves are boosted by the new programme; the programme is in the scallops' interest. Callon introduces the term 'obligatory passage point' (OPP) to describe the way in which the fisheries scientists construe and offer their scallop-rearing procedure as the only answer to everyone else's problems. If the fishers want to stay in business, they will have to channel themselves through the researchers' OPP. If scallops are to survive in St Brieuc Bay, they too will have to play along. And if other scientists want to have more knowledge about French scallops, then they too need to take advantage of these researchers' innovation.

But, of course, it is by no means certain that these others will succumb to the problematisation proposed by the researchers. The proposals impose costs on the other actors: fishers will have to give up some of their time, others will have to invest money and so on. They will have to forswear other proposals which may be put to them. Accordingly, Callon argues that the second moment of translation is a process he describes as 'interessement'.[2] Interessement is defined as: 'the group of actions by which an entity (here the three researchers) attempts to impose and stabilise the identity of the other actors it defines through its problematisation. Different devices are used to implement these actions' (1986: 207–8). Interessement is achieved by 'devices', techniques (so to speak) of ensnarement. The scallops, for example, are 'interessed' by confining them in fine netted bags. These bags provide the young scallops with somewhere sheltered to tether and with a through-flow of sea water to supply nutrients, but they are also intended to prevent the mollusc larvae from becoming dispersed. The interessement of the fishers proceeds in a different way, through repeated meetings with fishers' representatives at which the researchers ram home messages about dwindling scallop numbers and reminders about the Japanese experimentalists' successes. In sum, 'For all the groups involved, the interessement helps corner the entities to be enrolled. In addition, it attempts to interrupt all potential competing associations and to construct a system of alliances. Social structures comprising both social and natural entities are shaped and consolidated' (1986: 211).

This process of interessement leads to the possibility of enrolment (now given a more specific meaning than in the earlier paper with Law). Once enrolled, others become participants in the researchers' scheme. As Callon puts it, 'Interessement achieves enrolment if it is successful' (1986: 211). The other parties' concerns are translated into the project adopted by the researchers. Without changing their supposed interests, fishers and other new backers come to adopt a new approach. Their previous aims have been translated into the terms of the new proposal. The attainment of enrolment may require lengthy negotiations or it may be achieved quite easily. No longer an alien proposal associated with outsider researchers, the hatcheries proposal has been turned into a realisation of what was 'all along' in the other actors' interests.

The final moment of translation is termed by Callon the mobilisation of allies. If they have successfully enrolled the other actors in their project, the researchers can now aspire to behave as the spokespersons for the whole chain of allies: for the scallops, the fishing community and the scientific specialists interested in molluscan life. Callon's claim is that mobilisation allows the exercise of power over and through one's allies. The researchers from Brest can now speak for the other actors without constantly referring back to them; 'To speak for others is to first silence those in whose name we speak' (1986: 216). Furthermore, one's allies can be mobilised in a variety of ways: the scallops can be invoked through a graph plotting their numbers; fishers can be mobilised through data on their catches or incomes. Callon emphasises the connotations of the term 'mobilisation' since he wishes to suggest that one's enrolled allies are truly rendered mobile. The scallops stay in the sea, but graphs representing their numbers are much more mobile. The graphs can accompany a scientific paper or form part of a sales pitch to scallop salespersons in a Parisian market; Latour refers to mobilisations such as graphs and charts as 'immutable mobiles' (1987: 227). The *fruits de mer* can be fruitful far from the sea.

However, in the case selected by Callon, the story does not end happily for the alliance. After a while the scallops turn dissident; they seem to refuse to anchor in the devices installed by the researchers. The fishers defect also; growing impatient with the project, a group of fishermen dredges up all the scallops hatched in the early years of the programme and takes them to market to profit from the peak Christmas demand. Other scholars seem to become doubtful of the researchers' work; possibly the Japanese findings do not apply to the Atlantic varieties of scallop. Finally, the funding on which the work depends is called into question. The researchers must give up or start the whole process of enrolment all over again.

THE DISTINCTIVENESS OF ACTOR-NETWORK THEORY

Though the story of the scallop-rearing enterprise is not necessarily the kind of case that would appeal to interest theorists or proponents of

EPOR, there is nothing particularly controversial for them up to this point. One can readily imagine how a follower of Bloor or Collins might be drawn to a study concerning controversial claims about shellfish-farming techniques, where some actors claim that the method works and some deny it. For example, a great deal of controversy has surrounded the possible environmental consequences of salmon farming in north-west Europe, with prolonged disputes over both the environmental impact of faecal matter from the fish and the consequences of bulk medicines administered to the caged salmon. Given that salmon farmers have favoured sheltered marine locations on the Irish, Scottish and Norwegians coasts, and that these regions are also believed to be of significant wildlife value, these disputes have been trenchantly fought. Such disputes could easily become the focus of a controversy study. And a follower of Collins might even adopt some of the terms introduced in Callon's study (interessement or enrolment and so on) to describe moves made during such a controversy. However, it is with his treatment of symmetry that Callon stands out from the authors reviewed up to this point. In order to complete the move towards symmetrical analysis, Callon uses the same terminology of enrolment and so on to refer to the shellfish themselves as well as to the social actors. 'Problematization, interessement, enrolment, mobilization and dissidence ... are used for fishermen, for the scallops and for scientific colleagues. These terms are applied to all the actors without discrimination' (1986: 221). Given all the arguments about the correct way to approach the explanatory role of the natural world advanced by Bloor, Collins, Barnes and others, Callon makes this crucial analytic move in a surprisingly subdued manner; I shall return to this point in the next section. At this stage, however, I shall carry on with the elaboration of the Actor-Network Theory position.

The fundamental model of scientific activity that Callon and especially Latour put forward is that actors, whether individuals or institutions, whether Pasteur or the scallop researchers, attempt to build long chains of associates or allies. In these chains, the proponents of scallop farming and Pasteur try to secure for themselves pivotal positions as obligatory points of passage. Consequently, scientific controversies are effectively 'trials of strength' between competing alliances. In the St Brieuc case, the chains might have extended from the infant scallops, through fishers, the scientists and financial backers to the marketing and sales organisations which want to deliver scallops to Parisian gourmands. Building a robust and lengthy chain of associations is tantamount to winning the controversy. The analogy is a martial one: with a powerful enough chain of allies, one is more or less invincible. Instead of a controversy being decided by which 'side' has the best access to the truth, the central idea of ANT is that the truth results from building a successful alliance. The other distinctive thing about the ANT position is that these alliances are viewed as heterogeneous and open: they are composed of actors, and institutions, of technologies and non-human actors. According to Callon's case study, the scallop fishery ended unsuccessfully. But suppose it had succeeded and the parties involved had developed and marketed scallop-rearing kits; the existence of these kits on the market would have become part of their alliance. Somebody who wanted to argue with this new fisheries programme would now not only have to contest the scientists and the fishers, but deny the efficacy of the kits as well, and so on. Thus, the chains of allegiance are open in so far as new allies can be generated and 'recruited' as the alliance develops. In this sense, the kits themselves would become 'actors' in the controversy. Building novel entities and partners into well institutionalised and often automated procedures (such as these conjectural scallop-rearing kits) is referred to by Latour as black-boxing. He summarises his view as follows:

> We always feel it is important to decide *on the nature of the alliances*: are the elements human or non-human? Are they technical or scientific? Are they objective or subjective? Whereas the only question that really matters is the following: *is this new association weaker or stronger than that one*. Veterinary science had not the slightest relation with the biology done in laboratories when Pasteur began his study. This does not mean that this connection cannot be built. Through the establishment of a long list of allies, the tiny bacillus attenuated by the culture has a sudden bearing on the interests of farmers. Indeed, it is what definitively reverses the balance of power. Vets with all their science now have to pass through Pasteur's laboratory and borrow his vaccine as an incontrovertible black box. He has become indispensable. The fulfilment of the strategies [of translation] is entirely dependent on the new unexpected allies that have been *made to be relevant*. (1987: 127, original emphases)

A further analytical point can also be made at this stage: the key claim of ANT is not that the alliances are built up to override appeals to the truth. Successful alliances constitute the truth about whatever domain they are influential in. Latour has repeatedly been clear about this point: what appears afterwards to be simply an alliance with truth on its side was, in fact, the alliance that built the truth. Part of the work which scientists put in when constructing knowledge is the burying of the traces of this constructional activity (1987: 99).

Actor-Network Theory therefore sets out a distinctive approach to understanding knowledge in society. Scientists (as well as other kinds of actors) build chains of allies to take their projects forward. To build these chains, actors translate others' interests into their own (through problematisation, interessement, enrolment and mobilisation). They make themselves central to these alliances by presenting themselves as obligatory points of passage, indispensable steps in other agents' attainment of their own goals. Other elements in the chain can be mobilised, often in symbolic forms such as graphs, charts and images; these are the so-called immutable mobiles. The chains agents build are heterogeneous, consisting of people, things, devices, techniques, texts and symbols, and so on. Chains may be consolidated by black-boxing their components, for example by standardising a device (a measuring instrument or a statistical tool for example) so that it becomes harder to unpick or deconstruct. None the less, chains are as strong only as their weakest link (Latour, 1987: 121). Scientific controversies are trials of strength between competing alliances. The accepted truth, as well as the accepted black boxes and authorised devices, are the consequence – the outcome – of these trials of strength.

TWO LINES OF CRITIQUE

Actor-Network Theory has grown to be a highly popular approach within science studies. The innovative terminology it has introduced has been widely adopted. And its emphasis on the practical objectives implicit in knowledge-making (the recognition that scientists have to build alliances with funders and equipment suppliers if they are to advance the construction of knowledge) has proven an attractive alternative to the rather cerebral image of science advanced in most of the philosophical traditions, such as that of Popper or Lakatos. However, two principal forms of criticism have repeatedly been directed at the programme outlined by Callon and Latour (see Yearley, 1987 and Collins and Yearley, 1992a; also Amsterdamska, 1990; Schaffer, 1991). The first appears the more niggling of the two, though it turns out to have profound implications. One of the things about which ANT boasts is its ability to transcend social constructionism because it extends symmetry to all kinds of actors. It was this heterogeneity which Callon was eager to proclaim in his early case study and this was the core of his critique of Bloor's symmetry proposals. It is also why Latour now calls himself a constructionist but not a *social* constructionist.

The difficulty with this line of reasoning is that it begs the central question with which modern science studies started. In order to talk about the scallops or microbes or other elements in the natural world being enrolled in the network, one needs to know how they behave. Sociologists claim to know how social actors behave because they (or rather, we) study them. But how does Callon know about the behaviour of scallops? In practice, unless Callon has privately made extensive study of scallop behaviour himself, his account of their behaviour depends either on his common-sense assumptions about what scallops do or he has to take his lead from scientists or other human actors (maybe fishers) who do study scallops. Similarly, for Latour to talk about Pasteur enrolling microbes, one needs to know about the behaviour of microbes to see if they were truly acting in support of Pasteur's alliance (it should be recalled that in the quote cited above, Latour claims that it was the bacterium which 'definitively reverses the balance of power'). In practice, therefore, ANT becomes dependent on taking at face value the very scientific opinions whose success it is seeking to understand. For these reasons, Collins and Yearley (1992a: 314) assert that 'as a social account of the making of knowledge it [Callon's study] is prosaic, because the story of scallops themselves is an asymmetrical old-fashioned scientific story'.

This criticism is not entirely cut and dried. Actor-Network theorists might reply that they don't need to know comprehensive information about the behaviour of scallops, of Pasteur's bacteria and so on. They only need to know about the 'actions' of scallops on certain occasions when those actions are pivotal to the debate or about specific actions which the scallops manifest. Thus, if the scallops fail to anchor in the Brittany nurseries, Actor-Network analysts do not need to know why the scallops did not attach, they only need to note *that* they did not. It is not necessary to establish that scallops, so to speak, wholeheartedly sign up for the alliance, only that particular instances of scallop behaviour are reckoned as displaying enrolment (or dissent). Similarly with the case of Pasteur, one might argue that advocates of ANT do not need to detail everything about the behaviour of microbes. There is only the matter of whether mould is produced in a series of vessels which indicates the mould's 'allegiance' with Pasteur. On occasions, Latour seems willing to push this analysis for the case of human actors as well; it is the traces of human actions, not the whole actor, that are important. One needs to know whether the fishers organise themselves around the new possibility of scallop farming; one does not need to know everything the fishers do. The chains of allegiance are thus composed not of 'whole' actors but of what one might call 'act events'. In this sense the alliances are symmetrical in their composition: they are made up of traces of actors, what Latour has called actants, whether human, animal, bacterial or technological.

But for this manoeuvre to work, Actor-Network analysts still have to be able to determine what the scallops (and so on) did in fact do. This is not as simple as it sounds. Scientific controversies often turn on what the data truly were. In Pasteur's famous controversy with Pouchet over the spontaneous generation of life, for every experiment in which Pouchet found microscopic life in his trial vessels, Pasteur conducted one in which evidence of life did not appear. In this case, trying to step back from what the microbes are like in general to what the microbes did on some specific occasion does not work, because one cannot tell whether the microbes were acting typically or not on any particular occasion. Whether Pouchet's discovery of life or Pasteur's discovery of no-life was the better indication of what microbial life is like was not indicated by the tests alone. Similarly, in the gravity-wave case reported by Collins, the presence or absence of gravity waves was precisely the issue in dispute between the rival detecting groups. If one is going to enlist traces of actants in the alliances there is still the question of knowing which traces are robust enough to lend weight to the alliance and which are so questionable that they are likely to undermine any chain in which they are implicated since, as we have already seen, chains are only as strong as their weakest links.

Thus this first problem with the challenge to symmetry seems to run deep. To achieve the radical symmetry favoured by Callon, ANT has to be able to include natural agents (actants) in its chains of allegiance. But to work out whether those natural actors are dependable allies in the chain begs the very question the study set out to resolve. ANT wishes to argue symmetrically that Pasteur and his allies succeeded (in part) because he was able to enrol the microbes as well as human actors and organisations. But his successful enrolment of microscopic life depends on the correctness of his beliefs about that life, something that was established by the victory of his alliance.

Put this way, the first difficulty leads directly on to the second one, the problem of tautology. If scientific controversies are to be understood as trials of strength, then one needs some way of gauging the relative strength of the alliances. If the only proof of the strength of the alliance is the fact that it was victorious, then the whole procedure is manifestly circular. ANT can of course weaken its claim a little at this point. It may suggest that, by and large, in controversies the parties are well advised to build alliances. It can add the useful observation that alliances can be built of heterogeneous ingredients. Unlike philosophers of science who looked for the strength of beliefs only in their cognitive robustness, the idea that strength may be built out of multi-member allegiances is novel and helpful. But without further analysis of what the strength of an alliance means, the ANT 'theory' is simply hollow. Controversies are resolved in favour of the stronger alliance; the superior strength of that alliance is demonstrated by the fact that it won the contest. Such a theory is bound to be correct 100 per cent of the time, as well as hopelessly unilluminating.

SCIENCE STUDIES IN A NEW PLANE

Up to this point I have presented ANT as though it were straightforwardly a competitor to interest theory, EPOR and the traditional philosophy of science, trying to do the same job as they set themselves, namely explaining how it is that certain scientific beliefs, theories and practices come to prevail. There is much in the writings of ANT authors to support this interpretation (for example in Latour and Woolgar, 1979 and Latour, 1987) but there are also a few explicit claims to the contrary. In a number of writings from approximately 1990 onwards, Latour in particular has departed from this stance, developing an argument claiming that science studies should be operating in a new plane (see also Callon and Latour, 1992: 346); in 1999 he entertainingly claimed that there were only four things wrong with Actor-Network Theory to date, namely the understandings of 'actor', 'network', 'theory' and, for good measure, the hyphen (1999a: 15). On this more recent view, the empirical programme of relativism and other strong social constructionist projects stand at one end of a continuum, the other end of which is occupied by realist philosophies (this view was foreshadowed at the end of Chapter 2 when realism and constructivism were described as intimate enemies). These adversaries squabble furiously, but they only argue about how much or how little social construction there is. Latour claims that they are both flawed because they fail to rise above that one-dimensional squabble.

His rejection of this flat world is underlined by his repudiation of the term 'social construction'[3] in his most recent collection of essays, *Pandora's Hope* (1999b: 91), where he asserts that 'Science studies does *not* occupy a position inside the classical debate between internalist and externalist history. It entirely reconfigures the questions' (original emphasis). The best example of Latour exemplifying his own recent analytic approach comes from the first case study reprinted in *Pandora's Hope*. This rather well-known paper deals with Latour's relatively brief study of field scientists carrying out an analysis of the forest/savanna boundary in Brazil, where he aims to show that knowledge is not either the product of society *or* of nature but the outcome of multiple translations.

The researchers whom he is studying are trying to work out whether parts of the forest are advancing into the savanna or vice versa, or indeed whether the boundary line is stationary. Botanical evidence seemed to suggest that there are advance guards from the forest entering the savanna, though it is hard to keep track of the forest in the absence of long-term, accurate records. The situation is further complicated because, on the face of it, the soil beneath the habitat types is distinct: sandy below the savanna and clayey beneath the forest. Soil scientists' expectation is that clay can be degraded to sand, but not vice versa; sand cannot be 'upgraded' to clay. Accordingly, soil science suggests that the forest must be in retreat or, at best from the forest's point of view, at equilibrium. The team of field scientists that falls under Latour's scrutiny is trying to settle this question.

Latour acknowledges that a scientific audience would typically be interested in whether or not the forest is advancing. An ordinary sociology of science account by an interest theorist might be aimed at examining why these scientists or associated policy-makers or campaigners come to believe that the forest is or is not shrinking. But Latour wants to quit this quarrelsome plane. He claims to be interested in something distinct: how knowledge is constructed. His focus is on what this case shows about how the forest and savanna can be de-localised and brought back to research facilities elsewhere in Brazil, even back to France and, further still, into print. He aims to demonstrate how the practices and actions of these field scientists bridge the gulf between minds and things, not by reducing one entity to the other, and not through any one-shot technique. Rather, an artfully linked series of translations generates the shape of the forest/savanna boundary and allows representations of that boundary to circulate in print, as maps and charts. In a review article, Lynch has expressed the flavour of Latour's study very precisely:

> [In Latour's account] [t]here is no paradigmatic gaze, no single moment of discovery, no ultimate confrontation between an object and a theory- or concept-laden interpretation. Instead, there is an assemblage of interventions inscribed upon diverse materials which are temporally organised into an evidentiary chain, each link of which represents painstaking efforts to conserve, preserve, measure, encode, and assemble evidence of what was 'already there' in the wild terrain investigated. The forest–savannah boundary is eventually enclosed in Cartesian coordinates, but not in accordance with a Cartesian dualism. (Lynch, 2001: 225)

Rather as ethnomethodologists do (and here Lynch's sympathy for Latour's 'project' becomes clearly understandable, see Chapter 6), Latour chooses to focus on how scientists accomplish the work of bringing nature back from the field and into the laboratory, and thence into print. As he elsewhere expressed this, ANT

> was never a theory of what the social is made of. ... For us, ANT was simply another way of being faithful to the insights of ethnomethodology: actors know what they do and we have to learn from them not only what they do, but how and why they do it. It is *us*, the social scientists, who lack knowledge of what they do, and not *they* who are missing the explanation of why they are unwittingly manipulated by forces exterior to themselves and known to the social scientist's powerful gaze and methods. ... Far from being a theory of the social or even worse an explanation of what makes society exert pressure on actors, it always was, and this from its very inception, a very crude method to learn from the actors without imposing on them an *a priori* definition of their world-building capacities. (1999a: 19–20, original emphasis)

Understood in this way, ANT is not a theory, at least not a theory in sociology. It does not try to explain actors' beliefs. Instead it tries to illuminate how knowledge is built. Rather like Quine and Bloor, Latour wants to do away with conventional epistemology. His proposal is to replace it with an appreciation of how actors put knowledge together through translation. A soil sample from Brazil is translated into a representative of a soil type, which is then translated into a sketch map of soil distributions, which then can be transferred onto a formal map and circulated in the scientific papers and laboratories of the world. It is only through these myriad devices and translations that knowledge can be built up, so much so that Latour ends his study by urging: 'Let us rejoice in this long chain of transformations, this potentially endless sequence of mediators' (1999b: 79). Or, as Lynch puts it, 'Latour does not doubt that the end-product of this work tentatively reconstructs the gradual movement of the forest–savannah boundary. Instead of encouraging skepticism, he describes the production of conditions of felicity for the researchers' graphic representations' (2001: 225).

CONCLUDING REMARKS

Actor-Network Theory is widely treated as a theory in the sociology of science. This is easy to understand since the approach appears to offer numerous advantages to the social analyst of science. It proposes a rich vocabulary of terms to describe the stages in a scientific debate (interessement, translation and so on). It appears to complete the Strong Programme by offering a general symmetry. And, since it appears to assign a conspicuous explanatory role to the natural world too, it has attracted analysts of science who felt uneasy at the apparent extremism of the Strong Programme or of EPOR. For many, it afforded what was in practice a moderate (though, as argued above, ultimately incoherent) version of the Strong Programme, albeit one with apparently more radical (because they were more symmetrical) credentials. The irony was, that without much changing their practices, historians and other analysts of science could begin doing ANT just by carrying on their case studies as before while adopting the new terminology. ANT's followers used the terms, but usually with an explanatory end in mind, an orientation which Latour now claims was absent from the very start. Most ANT 'analyses' in the literature follow Callon's early methodological precepts in an unquestioning and very problematic fashion.[4]

This is not at all to imply that ANT's terminology is without value. The great majority of the terms illuminate repeated moves in scientific debates. But Latour and Callon have recognised that they are not doing sociology or philosophy of science in the way that their opponents and many of their followers are. They are right that ANT is not a sociology of science. But whether it is a worthwhile social theory at all will be considered again, alongside ethnomethodological studies of science, in Chapters 6 and 7 (see also Lynch, 2001: 230).

---

[1]It may be helpful to point out that the spelling of 'St Brieuc' is itself not stabilised, featuring even in later references by Callon and Latour as St Brieux. In his 1986 publication, Callon uses the 'c' version.

[2]Though only just about an English word, I have used the anglicised spelling of the term 'interessement'; in French, it is written *intéressement*.

[3]Even on his website (http://www.ensmp.fr/~latour/), Latour is careful to announce himself a constructionist but not a social constructionist. In the FAQs, he writes 'Is BL a social constructivist? The answer is not quite. The word was used in the first edition of *Laboratory Life* in 1979 and then dropped in the second edition. ... But if is not [*sic*] a social constructivist, BL is certainly a full blooded constructivist.' In his essay review (2001: 226) Lynch makes the playful suggestion that, with the development of Latour's thinking, further words will have to be dropped from the title in subsequent editions, leading to the book eventually having only a one-word title; he proposes *Factishes*.

[4]An unwillingness to cause widespread offence inhibits me from citing many specific studies here. However, a recent case that illustrates my point is an insightful analysis of the way in which farm effluents became 'constructed' as a significant cause of river pollution in England and Wales during the 1980s. The authors (Lowe et al., 1997) nail their analytic colours to the ANT mast-head but then make no significant use of the distinctive aspects of Actor-Network Theory in their analysis at all; for further details see my review (Yearley, 1999b) and the study itself. Lynch too appears to accept this point about the hollowness of the invoking of ANT in many case studies (1993: 108) though he limits his critical remarks to 'American' adopters of ANT.

5 Gender and Science Studies

THE DISTINCTIVENESS OF GENDER AS A THEME IN SCIENCE STUDIES

It is easy to see why scientific knowledge matters to analysts of gender. In principle, science offers to tell us how the natural world is. Scientific methods of investigation accordingly ought to be able tell us to what extent gender differences or particular gender attributes are natural. According to the received views about the character of science, scientific knowledge should properly be value neutral; in other words it should tell us how things are, irrespective of whether the nature of things is to our political and cultural taste or not. The resulting claims about the natural differences or similarities between people of various kinds can thus be said to 'naturalise' these differences or similarities since they present them as based in nature and therefore in some ways as beyond human choice. As outlined in the introductory chapter, this commitment to value neutrality can on occasions be as comforting to liberal and progressive interests as to more traditional values. For instance, if the overwhelming evidence from natural history in the nineteenth and twentieth centuries suggested that we humans were descended from ape-like ancestors, then this appeared to be an established fact however discomfiting it might be to religious authorities who interpreted the spiritual distinctiveness of humans as requiring that human beings had a quite separate origin from soul-free animals. The same considerations have applied in the case of gender, though the detailed consequences have been rather more complicated.

In the last three decades this naturalising aspect of scientific knowledge has indeed been the most important from the point of view of feminist studies, but it takes a little time to follow the complex implications that derive from this history. At the most general level, feminist analysts have tended to be sceptical about such naturalisation, principally because it has commonly been women's disadvantages or men's privileges which have been naturalised. From early 'scientific' studies a century or more ago which claimed to find women's brains ill-suited to rigorous analysis (see Tuana, 1989: vii for a quoted example), to much more recent sociobiological work suggesting that male mammal infidelity is widespread, evolutionarily comprehensible and pretty much incurable, the naturalising scales appear to have been tipped in men's favour (Hubbard, 1990: 96). Men may not exactly be from Mars, but they are naturally distinct, with different skills, appetites and gender interests (interests such as maintaining their typical economic and cultural advantages over women).

To many feminist analysts, such supposedly scientific findings have appeared dubious, dubious not merely because they naturalise women's disadvantages but because the studies on which the results are founded often appear weak. This may be because they are based on rather few, not terribly systematic studies, perhaps studies which might not be seen as sufficiently convincing had they not come to apparently commonsensical, stereotypical conclusions. It may be because the findings echo taken-for-granted assumptions a little too closely, suggesting that the findings may result from people jumping to familiar conclusions. And it may sometimes be because any direct similarities between animals' or other biological systems' behaviour and human cultural patterns are taken at face value while dissimilarities are played down. We shall return to these themes later on in relation to celebrated studies of the behaviours of the egg and sperm and of extrapolations from ape behaviour. However, it must be pointed out that it is not inevitable that the naturalisation should support ideas which are inimical to women's presumed gender interests. For example, the claims of mainstream science can also be adduced to support the idea of fundamental similarities in intelligence across genders. In cases where women have been institutionally disadvantaged because of supposed sex differences in intelligence, mainstream scientific findings about the absence of difference can be used to oppose discriminatory treatment. The naturalisation of similarity, at least in this case, appears welcome and progressive. Much feminist writing in the late 1960s and 1970s was directed at showing that in fact (so to speak) women and men were not as different as had generally been supposed and that women were not unfit for demanding occupations, nor men unfit for childcare. In 1972, for instance, Oakley approvingly wrote that 'biology also demonstrates the *identity* of male and female – their basic similarities, the continuity in their development' (1972: 18, original emphasis).

A key analytical question therefore is whether analysts of science should worry about all 'naturalisation' or only about erroneous or premature attempts to naturalise difference. One leading trend within feminist science studies has been towards a reformist agenda. This kind of approach (dubbed 'feminist empiricism' by Harding, 1991: 111) proposes that the solution is to have more careful science, which has been alerted to the dangers of too-hasty naturalising. The case made here is that the best antidote to false generalisations is to expose their shortcomings and to replace them with sound generalisations. Of course, as reviewed in

Chapter 1, the original Mertonian suggestion was that the scientific community had a core commitment to universalism so that the disciplined advance of science would likely be the best way of overcoming tendentious claims. To make sure that this occurs, various types of social reform of the scientific profession are advocated. Getting more women into the profession and into leading positions should provide people who are more conscious of the dangers of specious generalisations about female characteristics. One can accept the idea that scientists may be in danger of jumping to prejudicial conclusions even in the course of their scientific work (whether about gender characteristics or the nature of gay people or people from ethnic minority populations and so on) while still regarding more and better science as the way to overcome this danger.

However, this reformist position leaves at least two issues unaddressed. The first is the nature and source of the prejudicial judgements; the second is the possibility that there are limitations to the prospects for transcending these prejudices simply through trying harder with today's science. Concerning the first of these, feminist scholars face similar difficulties to those confronting the interest theorists described in Chapter 3. While it is easy to point to ways in which social institutions and beliefs appear organised in such a way as to favour men's interests (through, say, evolutionary psychologists' interpretation of men's infidelity as natural), it is hard to specify exactly how men's interests are co-ordinated to ensure these inequalities are legitimated and (thereby?) perpetuated. Followers of Marx, by contrast, had a clear (if erroneous) notion about what the supposed interests of the capitalist class, and therefore contrariwise the working class, must be. But when feminist scholars invoke the idea of patriarchy to comprehend how institutions and beliefs can be structured so as persistently to disadvantage women, it is difficult to work out what exactly the interests of men are, given that men are so diverse, with such conflicting interests themselves. Men appear to prolong their hegemonic control through subtle, pervasive and uncoordinated actions. One would need some well-developed account of why this hegemony persists before deciding whether feminist empiricism is a correct and sufficient remedy.

Second, one can find within the feminist science studies literature the idea that, given the chance, female scientists do (or would do) science differently; they might, for example, use different methods, employ alternative framing assumptions, or approach the topic more holistically (for an empirical investigation of this issue see Kerr, 2001). If this view is right, then culturally oppressive science is not to be corrected by more of the same, but by different and better approaches. But the analytic danger here is that this threatens to exempt feminist science from sociological study altogether. The sociology of 'malestream' science amounts to critique and exposé while the study of feminist science equates with celebration.

Moreover, there is a well-recognised danger here of essentialism. If one wishes to make the argument that women would do science differently, one needs to say what it is about women that sets them apart epistemologically. The simplest argument here would be that it is women's 'femi-nineness' that distinguishes them. But this route is obviously unattractive since it requires that all women (across all cultures and time) have something in common and something different from men. Worse still, even if one could find a candidate for what this special qualification might be, it is very uncertain how any special characteristics that one could readily imagine ascribing to all women could shape the way that women would undertake the sciences. Even those critical of feminist empiricism, including Harding (1991: 121), are keen to avoid the pitfalls of essentialism. The favoured route out of this impasse has usually been labelled as 'standpoint' theory, an approach I shall review later on. At this stage it is helpful to appreciate that feminist scholars are aware, as Longino points out (1990: 11), that there is a danger of trying to have it both ways: arguing that apparently sexist scientific findings are empirically false (which implies they might be correct) and arguing that they are wrong in principle (in which case, so are the empirical disproofs as well as any empirical support for more egalitarian outcomes). Before proceeding further with abstract, theoretical arguments, it will be helpful to review two cases which allow us to see how these arguments work out in practice.

MANLY SPERM AND 'GIRLY' EGGS

The phenomenon of naturalisation can be helpfully exemplified through a (much-reprinted) study conducted by Martin (1996) based largely on the sections of a range of biology and medical textbooks that deal with human reproduction. Unsurprisingly, these texts report that women's bodies carry eggs and men's supply sperm. But the accounts of how the eggs and sperm behave tend to present eggs as stereotypically feminine and sperm as masculine in human (indeed Western) terms. Thus the egg is presented as passive; it is swept along the fallopian tube. Sperm are active, propelled by strong tails in a competitive race to reach the egg (1996: 327). Furthermore, it is the sperm that are textually presented as the more active agents in the process of fertilisation itself: they penetrate the egg, burrowing through the coat of the egg and 'activate the developmental program of the egg'. Sperm are presented as more independent of the context than are eggs, even though both have limited 'lives' unless fertilisation takes place. Martin even finds one text which claims that, 'To execute the decision to abandon the haploid [the unpaired] state, sperm swim to an egg and there acquire the ability to effect membrane fusion' (cited in Martin, 1996: 329).

Martin draws attention to the corporate tone of this description. The clear point here is that the female reproductive elements are given stereotypically demure, feminine characteristics, while the sperm behave like competitive young men looking for promotion in blue-chip companies. As Hubbard notes (1990: 102): 'It reflects the ideology of gender relations, in which males pursue and females yield.' The irony of course is that eggs are not themselves female in any obvious sense; eggs do not have a sex, even if they are biologically associated with females. There is no reason at all to expect them to manifest feminine attributes.

Martin goes on to consider some more recent research which casts doubt on the received views of the egg and sperm. Some measurements in biophysics suggested that the movements of sperm, though strenuous, are not well designed for penetrating the egg at all. During sperm 'swimming', the head is propelled from side to side and is thus not likely to force its way into the egg in any event; if anything the characteristic movement of sperm would tend to shake them loose from any egg they encountered. This cast doubt on the idea that penetration could be a physical phenomenon, a physical puncturing of the egg coat, and even made it appear unlikely that sperm could burrow into the egg using chemical means to pierce the egg's outer layer. Instead it was proposed that the egg and the sperm become fastened 'because of adhesive molecules on the surfaces of each' (1996: 331). According to Martin, this means that 'the egg traps the sperm'. Other biological researchers who adopted similar views none the less managed to couch the encounter in stereotypical terms. One proposal was that the sperm and egg bind, thanks to a filament which is constructed out of protein stored in the sperm. The authors described this filament as a device for 'harpooning' the egg, even though – unlike a harpoon – this filament only sticks to the surface of its quarry (and does not penetrate it) and is built up out of sections, rather like a pontoon bridge (and is thus not much like a projectile).

Furthermore, even when authors treat the initial adhesion between the egg and the sperm as an interactive process, there is still a tendency to talk of the progress of the sperm into the interior of the egg to meet the nucleus as an act of penetration, even though the same arguments about the weak propulsion of the sperm apply here too. Research results from studies of certain species (Martin lists mice and sea urchins) imply that sperm may lose all motility on fusing with the surface of the egg. In the sea urchin case, for example, the sperm is drawn into the interior of the egg by microvilli (tiny 'hairs') developing from, and withdrawing into, the nucleus. Yet in the text giving an overview of this work the section is headed 'Sperm Penetration', even though – as Martin points out – it could as readily be entitled 'The Egg Envelops'. And on those occasions when the egg is accorded a more active role in the text, Martin detects that the accounts 'bring into play another cultural stereotype: woman as a dangerous and aggressive threat' (1996: 336). As soon as agency is granted to the woman's egg, the egg begins to act as a 'femme fatale', luring the sperm into a sticky, spider's-web-like trap.

Finally, Martin points out that this stereotypification extends beyond the imputed characteristics of the egg and sperm. According to her analysis, medical and biological texts tend to treat the reproductive abilities of the two genders in culturally loaded ways. The ability of men to 'manufacture several hundred million sperm per day' is presented as a prodigious talent whereas 'the newborn female already has all the germ cells she will ever have' (1996: 324 and 325). Women have a stock of reproductive material which they merely use up while men are wonders of creativity. Even the verbs selected reflect this ranking of men and women's attributes: men 'manufacture' sperm while women 'shed' eggs. And the way that women and men 'manage' their reproductive assets can be treated in a similar manner. According to Martin, the fact that women start life with more eggs than they are able to use in their monthly despatch, is treated as wasteful, whereas the enormous disproportion between the number of sperm produced and those that achieve reproductive success is presented as a testimony to the creative power of men. Sperm seem to be valued whether they achieve reproductive success or not; unfertilised eggs are seen as worthless, a stockpile already approaching its sell-by date. And this seems to reflect, even 'justify' and perpetuate, aspects of the inequality between men and women.

Martin is clear in claiming that these linguistic stereotypes are pernicious: she asserts that presenting feminine eggs as femmes fatales is 'damaging' (1996: 337) and proposes more neutral, cybernetic metaphors for use instead. But it is not wholly clear where exactly the alleged dangers are supposed to lie. Her study does demonstrate that scientific language is not free of metaphor and that this metaphor can align with conservative social assumptions. These biological systems, which are not obviously male or female in the biological, let alone the contemporary Western cultural sense, are interpreted through the lens of cultural expectations of women's and men's behaviour. If sperm are interpreted as behaving in typically male ways (even when other equally plausible interpretations would be available), and if this male conduct is then offered as evidence of how profoundly natural it is for men to behave in these ways, the naturalisation appears extremely suspect and hollow. But Martin does not show that these texts are in fact used to naturalise stereotypical male behaviour. Of course, readers of the texts may have their prejudices subtly confirmed or they may absorb these cultural messages along with the biological instruction. But Martin's analysis is wholly a textual one and thus tells us nothing about actual readers, except for Martin herself who – as a reader – is evidently quite able to reject the implied stereotypes. There are also two concerns about her selection of materials for analysis. Her paper is written as an interpretative essay, with little attention paid to the representativeness of the texts or the excerpts which she selects. Much more importantly, since the majority of the writings she selects come from textbooks, it is not clear that these covert assumptions affect the development of scientific knowledge at the forefront of research. Her findings can be read as a valid critique of stereotyping in science education without that necessarily having any implications for how novel science gets to be constructed. Indeed, the fact that she uses recent research findings (on the microvilli in mice and sea urchin ova, for example) as the basis of some of her critique of the textbook accounts, could be read as suggesting that innovative science is able to debunk the older sexist assumptions just as feminist empiricism would suggest. In sum, Martin's account illustrates how naturalisation can arise through the imposition of cultural metaphors onto biological phenomena. But it neither demonstrates that this naturalisation has an ideological impact on readers nor shows that naturalising assumptions influence the development of scientific knowledge on the frontiers of research. However, this critical issue of the link between naturalisation and developments at the forefront of scientific knowledge can be investigated through the next case study, a well-known example explored both by Hubbard and by Longino.

DO WOMEN OR MEN LEAD HUMAN EVOLUTION?

The outline steps in the history of human evolution are broadly agreed. Ape-like creatures (hominids) moved from the trees into the grasslands, began to have an upright bearing, changed their characteristic diet and developed extensive tool use. Their dental patterns changed, allowing the creatures to take advantage of new forms of food; there was also considerable cerebral and intellectual development. However, these processes occurred gradually and few traces remain to document these developments. It is difficult, therefore, for scientists to date the key transitions (the adoption of an upright stance for example) or even to put them in order. None the less, given elements of data – finds of teeth worn in characteristic ways, footprints, discarded tools and so on – together with a certain amount of inference, scientists have been able to piece together a story of hominoid development, a theoretical interpretation which draws together and makes sense of the data. Longino points out that the version which has achieved the widest circulation is a story of 'man-the-hunter'. Longino's study aims to show that an alternative woman-centred (gynecentric) account is at least as plausible but has not been offered in the same way.

Both 'man-the-hunter' and 'woman-the-gatherer' theories of human social and anatomical evolution place one sex's changing behaviour at the centre of the species' evolution though 'neither assumption is apparent from the fossil record or dictated by principles of evolutionary theory' (1990: 107).

Tool use is central to both accounts, since those hominids that used tools would have enjoyed clear adaptive advantages. At the same time, it makes sense that hominids that adopted a two-legged stance would be in a favourable situation since they would free their 'arms' and 'hands' to manipulate the tools; the same goes for standing upright. Hominids that developed 'characteristic human forms of intelligence' would also be at an advantage in the sophistication of their tool use (1990: 107). These changes could come to be linked in a virtuous circle; the more upright and smarter the hominids (up to a point at least), the better their tool use. In the male-centred account, 'the development of tool use is understood to be a consequence of the development of hunting by males' (1990: 107). On this androcentric view, it is in the course of the primarily male activity of hunting that hominids evolve the use of tools, setting the other evolutionary changes in motion. The availability of fighting tools means that canine teeth, larger in males and used for display and for actual fighting, become less important. The teeth can shrink, allowing the molars to be used more effectively in grinding food down (previously the canines would have got in the way).

Longino's conclusion follows readily from this outline of the androcentric theory. Males are at the centre of evolution, since this account

> ties the behavioral changes that contribute to selective pressures favoring the development of hominid morphological characteristics to male behavior. And not just any male behavior but behavior that, still in the twentieth-century mind, epitomizes the masculine. (1990: 107)

Not only is it the actions of men which are at the centre of the story. But the very behaviour which drives evolution in the 'progressive' direction of tool use and enlarged intellectual capability is quintessentially male. Hominids reap a big evolutionary reward for having the male group members behave in accordance with present-day stereotypes of maleness.

The gynecentric proposal, by contrast, looks for the explanation of the rise of tool use to changes in females' behaviour. A shift from the forests to the less-productive grasslands exerted pressure on food collection, which conferred advantages on those who used tools to assist food-gathering. This pressure was exerted most conspicuously on females who had to feed 'not just themselves but their young through pregnancy, lactation, and beyond' (1990: 108). On this interpretation, tools were used opportunistically, both to garner food and for defence against aggressive animals. Females could have started to improvise with tools such as sticks or reeds, tools which would leave no remains for modern-day scientific analysis since they would rot down and decompose. This would date the inception of tool use rather earlier than many versions of the androcentric scenario which focus on the first evidence of stone tools; Longino proposes that this earlier date would fit better with current estimates of the timing of dentitional changes. Finally, on the gynecentric view, male dental changes may be put down to selection pressures from the females, who favoured the less aggressive, more sociable, 'dentally-challenged' males as partners.

It is important to point out that, in presenting this study in these terms, Longino is not claiming that the latter option is necessarily correct. Rather, she wishes to stress that both accounts are compatible with the known data and that both depend on assumptions, assumptions which often pass unacknowledged. For instance:

> critics noted early on the tendency of researchers to rely on male informants, to ask questions reflecting male preoccupations, and to pick as models societies that supported their conclusions – to use perceived aggressiveness in male baboons, for example, as a model of aggressiveness in male humans [even though chimpanzees – an equally obvious choice – are relatively more sociable]. (1990: 106)

Thus, conventional scientific work on the history of human evolution tends to prioritise the activities of males even though (according to Longino and to Hubbard) there is nothing in the data or in established theory which means that the evolutionary changes are any more likely to be attributable to males than females. Male analysts appear to have jumped to conclusions which seem to them obvious, even if those conclusions look very problematic from a feminist point of view. In this case, therefore, we do have plausible evidence that the construction of scientific knowledge at the forefront of research (and not just in textbooks or educational material) has been influenced by gendered assumptions. Such 'findings' threaten to lend (at least) covert support to the idea that males are the more significant gender and thus serve to naturalise male supremacy and gender inequalities.

ASSESSING FEMINIST EXPLANATIONS IN SCIENCE STUDIES

On the face of it, Longino (on the one hand) and Bloor, Collins, MacKenzie and so on (on the other) are engaged in very similar exercises: they are examining rival scientific interpretations and looking for the sociological aspects of the competing views and of those views' supporters. However, this similarity is in important ways only superficial.[1] Where Bloor and Collins are committed to symmetry and impartiality, Longino is avowedly asymmetrical. Her aim is to show that the androcentric account is less secure than its proponents appear to think – because it is based on unquestioned, gendered assumptions – and that a rival interpretation is more worthy of attention than the established scientific community appears to acknowledge. Thus, unlike 'mainstream' science studiers, Longino aims to be both an analyst of the social context of knowledge production and a participant in the evaluation of the resulting claims to knowledge. The principal question therefore is how successful is Longino in defending and carrying out this joint role. Bloor's and Collins' commitments to symmetry and impartiality lead them to abjure such joint roles; they doubt whether a joint undertaking is even possible. From their perspective Longino often appears to be 'simply' a practitioner in the study of human descent. Of course, from the point of view of EPOR, in so far as Longino is engaged in feminist empiricism, then this is unproblematic. Realism is the natural attitude within the scientific community and if Longino is engaging as a participant in the science of human origins then an attitude of philosophical realism is straightforwardly appropriate. But, at some points in her analysis, she seems to adopt a position at odds with feminist empiricism, a position which centrally involves an argument about values in science of the sort discussed in Chapter 1 in conjunction with the work of Kuhn and Newton-Smith.

Longino argues that scientific claims are indeed evaluated in terms of what she dubs constitutive values (1989: 206), values which she does not list but which appear to be the kinds of cognitive orientations favoured by Kuhn and by Newton-Smith (to do with accuracy, consistency and so on). However, she asserts that these values will not themselves always suffice to direct scientific reasoning. Sometimes they may be sufficient but on many occasions they will not be, and other considerations will then have to play a role. Among these other considerations can be 'contextual values', that is, values stemming from 'the social and cultural context in which science is done' (1989: 206). The case of human origins appears to be (at least in Longino's view) an example of this kind of science. The history cannot be resolved by reference to the constitutive values alone because both narratives are compatible with the available evidence. Her proposal is that:

> Accepting the relevance to our practice as scientists of our political commitments does not imply simple and crude impositions of those ideas onto the corner of the natural world under study. If we recognize, however, that knowledge is shaped by the assumptions, values and interests of a culture and that, within limits, one can choose one's culture, then it's clear that as scientists/theorists we have a choice. (1989: 212)

In the later study from which the hominoid evolution case is taken she makes a similar point, arguing that 'feminist scientific practice admits political considerations as relevant constraints on reasoning' (1990: 213). In other words, when constitutive values fail to determine the outcome of scientific interpretation, one must turn to other values to achieve interpretative 'closure'; interestingly, this argument parallels Stage Two of Collins' EPOR programme. On such occasions, the scientific establishment has generally favoured conservative social values while feminists should (she believes) choose feminist ones.

It appears, therefore, that Longino's recommended stance amounts to the adoption of feminist empiricism in so far as particular bits of science can be directed by constitutive values alone. Where science cannot progress in this way, a feminist practice will lead feminists to choose the interpretation which best accords with their own cultural and political preferences.

But this analytic position is not as coherent as Longino appears to believe. The principal difficulty is that Longino is insufficiently sociological in her understanding of how values work. She treats the constitutive values in an almost wholly philosophical manner, overlooking the kinds of weaknesses already identified in the work of Kuhn and Newton-Smith and missing the extent to which cultural factors can influence the actual interpretation of such cognitive values. By contrast, in both MacKenzie's study of the measurement of association and Collins' study of cultures of proving in the gravity-wave community, it was clear that social and cultural differences led to varying interpretations of such supposedly constitutive values as 'conformity with existing statistical theory' and 'replicability'. In a sense, one of the main findings arising from MacKenzie's and Collins' case studies was that cultural and social factors influence the interpretation of scientists' 'scientific' values, not that social factors only come into play when the constitutive values have failed to deliver an objective result.

This neglect of sociological insights into the interpretation of values dovetails with the fact that Longino's study of the construction of histories of human evolution is offered in a very un-sociological manner. She relates scientists' views but says next to nothing about the context in which those views developed (save that it was within the male establishment); the reader does not learn why these scientists were interested in such questions nor understand what might follow from the answers. Adopting a more sociological understanding of the way that values are interpreted would allow Longino to give a more analytically consistent account of her case studies and to dissolve, at least partially, the rigid distinction she makes between constitutive and contextual values. This, in turn, would lead her away from an implausible dichotomy between cases where one needs only more and better science and other cases where one needs to adopt explicit political guidelines for making choices between scientific theories. It would lead her into the study of the contextual interpretation of scientific values.

Having identified weaknesses within the stance adopted by Longino, this leaves only the standpoint theorists' alternative unexamined. The standpoint argument develops in an Hegelian manner. It asserts that, at particular historical stages, certain sections of society are unable to see that societal arrangements could be other than they are. Typically, those who benefit from current cultural patterns view those patterns as unalterable, even literally unquestionable. Those in outsider positions may experience the culture differently and thus be able to call it into question. As mentioned in Chapter 3, Lukács advanced a Marxist variant of this argument, claiming that the working class had a distinctive capacity to develop an understanding of capitalist society. Harding and Hartsock develop this argument for feminists, with Hartsock suggesting that, 'A standpoint, however, carries with it the contention that there are some perspectives on society from which, however well-intentioned one may be, the real relations of humans with each other and with the natural world are not visible' (1983: 285). For example, in patriarchal societies men may see knowledge which underwrites their advantages as simply factual, whereas women, by virtue of their exclusion and inferior status, may be able to call these generalisations into question (see Harding, 1986: 155–8).

One can see how Hartsock and Harding believe that standpoint views escape the perils of essentialism since the point of view ascribed to women is not that of women-for-all-time but simply women under patriarchy. However, Longino none the less asserts that standpoint theories 'suffer from a suspect universalization' (1989: 205) since it is still not clear that, under patriarchy, all women do indeed share one standpoint. More fatally still, Harding and Hartsock fail to make the argument that the putative feminist standpoint has epistemological implications in any cases other than those already raised by Longino, Martin, Hubbard and other associated commentators. This is significant for two reasons. First, it suggests that standpoint theory is far from indispensable since commentators such as Longino did not need standpoint theory in order to make their claims in the first place. Second, it implies that the feminist 'standpoint' may have implications only for a limited sub-set of the subjects of scientific investigation. Despite Harding's ambitious assertion that 'it is the objective perspective *from women's lives* that gives legitimacy to feminist knowledge, according to standpoint theorists' (1991: 167, original emphasis), she fails to provide any evidence that women's alleged standpoint has any implications for scientific knowledge except in those (admittedly quite numerous) cases where scientists make assumptions about men's or women's behaviours, gender characteristics and so on, or ascribe patterns based on human gender stereotypes to the natural world.

CONCLUDING REMARKS

To try to summarise the avowedly feminist contribution to science studies, it is plain that feminist scholars have convincingly offered examples of the ways in which scientific knowledge dubiously naturalises gender differences and inequalities. Human gender characteristics are ascribed to bits of the natural world (such as ova) that have no gender. The history of human evolution appears to have been viewed through androcentric eyes and the resulting conclusions have been 'justified' by drawing parallels with other primate societies which happen to conform (approximately) to current human gender stereotypes. These tendencies appear to be pervasive. Important though such findings are, they do not demonstrate that gender issues are the most significant cultural influences on the shaping of scientific knowledge. Indeed, in Schiebinger's ambitious recent book-length attempt to ask whether feminism has changed science, she is reduced to offering two of Longino's cases and one other as her candidate evidence for changes wrought by feminism in 'the content of human knowledge' (1999: 181; see also the same examples employed for a broader audience in Begley, 2001). Feminist critiques – especially from within 'feminist empiricism' – have had an effect where scientific knowledge has naturalised gender differences, but the impact elsewhere has been (and appears likely to remain) slight. Furthermore, the prospects for the broad applicability of standpoint theories appear dim. This suggests, as I proposed above, that Longino's analysis is the most successful theoretical attempt to go beyond feminist empiricism. But its weaknesses (particularly in regard to the understanding of how scientific values operate) mean that supporters of this kind of approach would benefit from greater co-operation with advocates of the broad constructionist programme reviewed in the preceding chapters. Her analysis, though the strongest of the feminist theoretical positions, would be improved by being less formally philosophical in its understanding of the values underlying scientists' interpretation of data, evidence and results. The values guiding scientific reasoning are open to greater sociological variation than her analyses acknowledge.

---

[1]The superficiality of this similarity is also reflected in the extent to which the two literatures pass each other by. The sociological authors reviewed in Chapters 2–4 cite feminist scholars *very* infrequently. For their part, authors such as Longino

and Harding do cite some of the Strong Programme authors but engage with them rather little. Longino (1990: 10), for example, introduces the Strong Programme with a reference to feminist authors who had written about it and says little about the specific views of Barnes or Bloor. Harding relegates to a footnote the observation that Strong Programme 'sociology of knowledge is flawed in a number of ways' (1991: 167), but says no more about its alleged inadequacies. On the regrettable implications of this mutual inattention see Delamont 1987. Longino's more recent work (2002) does devotes more attention to science studies authors.

6 *Ethnomethodology and the Analysis of Scientific Discourse*

ETHNOMETHODOLOGY AS A SOCIOLOGY OF SCIENCE

Ethnomethodology and the analysis of scientific discourse are approaches to the social analysis of science which share one critical thing in common: they consider that science studies, as described in the preceding chapters, and particularly the sociology of scientific knowledge, are fundamentally misconceived. Both are concerned with remedying, or with providing better alternatives to, the flawed programme of science studies and both see the only hope for progress in the study of the fine details of what scientists write, do and say. They are also in agreement in so far as both are inclined to suggest that EPOR, the Edinburgh School, standpoint theorists and others overlook some of the particulars of science and tend to reinforce their own analytical perspective by playing down, or failing to attend to, the specific character of the scientific lifeworld. Yet, in many other ways they are not natural allies. Authors in the two approaches do not employ each other's work extensively; in fact, they tend to view each other as somewhat wrong-headed. Given their differences, it will be helpful to consider the approaches separately in the principal sections of this chapter.

There are many statements by ethnomethodologists outlining what they see their programme as offering or being about (among the more recent, see Garfinkel, 1996). But a common definition would suggest that ethnomethodology is the systematic exploration of the techniques used by members of a society to perform the tasks that make up that society's life. Ethnomethodologists are interested both in the ways that actors' techniques are adequate for carrying out the tasks of everyday life *and* in the ways in which the performance of those tasks is seen and acknowledged as adequate within the appropriate social milieu. Lynch et al. (1983: 206) assert that 'the overriding preoccupation in ethnomethodological studies is with the detailed and observable practices which make up the incarnate production of ordinary social facts'. By 'ordinary social facts' they mean such things as the orderliness of a queue or the regularities of turn-taking in conversation. By 'incarnate production' they refer to the way in which the orderliness of social life is accomplished in the very moment of acting. Using this standard as their benchmark, ethnomethodological writers characteristically argue that the sociology of science (as manifest in the preceding chapters) is not in practice a study *of* science at all. Science studies may be about science in some general sense but it is not a detailed empirical study of the doing of science, of its incarnate production. For ethnomethodologists, only ethnomethodological studies are truly studies *of* science. In this light Livingston has, for example, recently examined the activity of offering mathematical proofs (1999). He is adamant that he is not addressing Bloor's question about the nature of mathematical knowledge; he is apparently agnostic about whether mathematics details unalterable truths or whether it depends on contingent social conventions. Instead, he is interested in proving 'as a cultural activity' (1999: 867). He is interested in understanding, documenting and specifying what it is adequately to offer a proof in mathematicians' culture. As Livingston summarises it, one of his observations is that:

> When a prover stands before other provers and proves a theorem, the prover does not literally present a mathematical proof; the prover engages in the arts of description. The prover describes a proof of a theorem, as if that achievement were already in hand, and other provers attend to the prover's work at the blackboard as a description of that proof. ... mathematical arguments, in actual practice, are presented as descriptions of proofs that existed prior to their presentation. (1999: 873)

In other words, Livingston's study of the activity of proving shows that when a 'proof' is offered, it is offered as a description of a proof that already exists and this already existing status is attended to both by the person doing the proving and by those to whom the proof is offered.

It is important to point out that this position is not advanced by Livingston as a deduction of what a proof must be like; it is not meant as a transcendental deduction of the sort proposed by realists, including Bhaskar, and reviewed in Chapter 1. Rather, it is offered as a result of detailed empirical analyses of the features of proving as conducted in mathematicians' culture. None the less, this exploration of what the work of offering a proof looks like within the culture of mathematics does appear in some ways at odds with constructivist approaches which focus on issues such as the socially negotiated character of proofs. Livingston is interested in documenting and specifying what proof-offering is taken to be by mathematicians; of late ethnomethodologists have come to call this point of focus the haecceity of mathematical proving, where haecceity means 'the "just thisness" of any particular object or activity' (Lynch 1993: 283).[1]

This leads to the following kind of defence of the ethnomethodological position, as offered by Sharrock and Anderson in their overview of ethnomethodology and its relationship to disputes concerning the philosophy of scientific knowledge:

> Ethnomethodology need not step up to defend the conception of the reality of the science's phenomena in the sense in which the sociology of scientific knowledge typically challenges this. The issue is not whether scientists are right or wrong to hold 'realist' conceptions of their work, but whether the *fundamental* sense in which scientists find 'the reality of the phenomena' has anything to do with holding realist views at all. The question whether scientists are right in their 'realistic' construal of their achievements gives way to the question of whether the *scientists' sense of the reality of the phenomena* they deal with has in fact been identified at all. Ethnomethodology prefers to look into the ways in which scientists *encounter* their phenomena, to examine the ways in which they 'come upon these' in the course of their investigations, to see how – for example – their activities in a laboratory comprise – *as far as the scientists are concerned* – the disclosure of a hitherto undiscovered phenomenon (or, alternatively, the routine reproduction of a well established one). (1991: 74, original emphases)

This emphasis on how scientists' activities simply are (and are accountably seen as) the disclosure of an undiscovered phenomenon or the routine reproduction of a known one is made very clear in a study by Bjelic and Lynch (1992) of Newton's and Goethe's work on prismatic colour. By the start of the eighteenth century, Newton's corpuscular theory of light had become the dominant scientific interpretation, for example in relation to lenses and to understanding why it was that white light could be broken down into different colours. Goethe sought to challenge his views, particularly in relation to the analysis of colours. Bjelic and Lynch offer an unusual text relating to this historic debate, a text written as an exercise for the reader who, armed (ideally) with a modestly sized prism, can use the text and prism to produce the phenomena of which Bjelic, Goethe, Lynch and Newton write. Bjelic and Lynch's objective is not to review the history of the debate or to explain Goethe's relative lack of success in attempting to rebut Newton's approach. Instead, their avowed aim is to 'make perspicuous the embodied work of a (scientific) demonstration' and to do this by offering an 'installation of materials *in and as* the demonstration' (1992: 53, original emphasis; see also Bjelic, 1992). This cumbersome phrase, 'in and as', lies at the heart of what interests them most; the materials are simultaneously an account of the demonstration and comprise the demonstration itself. Their text aims to exemplify the point made also by Livingston when he observed that a mathematician's work of standing at a blackboard is simultaneously to offer a proof and to do the proving.

In a related vein, Button and Sharrock (1995; see also 1998) have made a study of how computer programmers write and learn to write code. They claim that programmers are schooled in ways of writing code that make it accessible to human readers as well, of course, as to the machines which execute it. Novice programmers are taught that human readability is key while the ease for machines is only 'incidental' (1995: 233). But how do programmers fulfil this obligation to make code readily understood? The authors' answer is that:

> The visual organization of the [well-written] piece has been designed to reflect the computational organization and to make *that* organization accountable. Computer programmers use visual *organization as an account of the computational organization of the program*. Through the visual representation of the computational organization, programmers are able to make the computational organization, or some aspects of the organization of the program, available for seeing. (1995: 248, original emphases)

They later go on to make this point even more explicit, arguing that the way the programme is visually organised on the code-writer's computer screen and on paper 'is an account of the computational organization in as much as it is designed' to coincide with that organisation (1995: 249). Button and Sharrock aimed to produce a study *of* program-writing, indicating how understandable code (understandable to human code-writers and to machines) is incarnately produced.

The three studies I have briefly outlined give a sense of what ethnomethodological analyses of science are like and what form their results may take. As Button has insisted (1991: 6), ethnomethodology aims to 'respecify' sociological studies. It aims not to be a new theory or school within sociology, but to make sociology over. It aims not to explain what mathematical proving is nor to explain why some particular mathematical proof was accepted when another was denied, but to show what the activity of proving is like, to elucidate the haecceity of proving. Similarly, it wants to document how optical phenomena are produced, to get at the 'just thisness' of using prisms to see spectra and to carry out experiments on them. And it seeks to show how computer code is produced as legible for the programming community. However, though these three studies may adequately impart a sense of ethnomethodologists' aims, they do not produce much in the way of 'results'. Bjelic and Lynch's textual 'installation' gives the reader the opportunity to 'use the instructions, equipment, and figures at hand in order to elucidate a phenomenon' (1992: 74) and in that sense allows the reader to see what the lifeworld of experimental science is like. But that is as far as it goes since 'to an extent, our project has a negative … consequence, deepening our initial suspicions about all attempts to posit science, scientific talk, scientific work, or scientific action as analytic objects for a general social science' (1992: 74). By contrast, part of Button and Sharrock's conclusion appears to afford a very straightforward, concrete finding: we learn that the visual organisation of code reflects (and is designed and seen to reflect) how the code is organised. But this appears to be an extremely low-level and unsurprising conclusion, a rather meagre version of the 'just thisness' of computer programming. Either way, the fruit of this respecification appears extremely modest.

The suspicious reader may suppose that I have chosen these studies precisely because it is difficult to derive informative conclusions from them. This is not the case; the three publications described above are among the more specific and elaborate cases available in the ethnomethodological literature on science and associated topics.[2] However, one further study which adopts a slightly different analytical orientation from within the ethnomethodological tradition provides a clearer indication of what the conceptual gains could be.

SCIENCE STUDIES AND DELUSIONAL TALK

In a recent study, Palmer (2000) has examined the sociology *of* symptom recognition in psychiatry. For around 30 years, sociologists have been debating with the psychiatric profession over the extent to which 'mental illness' is a social and medical construction. Whereas the medical profession has typically suggested that psychiatric malaises are based on biological and biochemical disorders, many sociologists have claimed that these malaises are – at least in large part – the result of social processes of marginalisation and labelling, and of ordinary people being placed in positions of intolerably conflicting obligations. On the face of it, this debate reproduces the divide between constructionists and realists concerning the status of scientific knowledge. Palmer's approach is to sidestep the debate by looking at the testimony of delusional patients and to ask the ethnomethodological question about how it is that psychiatrists 'hear' in the talk of disturbed patients the evidence of the patients' illness.

Delusions (persistent unsupported beliefs) have long been recognised as a symptom of certain sorts of mental indisposition. Palmer cites evidence from psychiatric manuals to show that psychiatrists' own definitions of how delusional disorders are to be recognised tend to focus on aspects of delusional beliefs themselves: such beliefs are said to be false, to be espoused with great subjective certainty and to be held irrespective of the evidence available (2000: 664). However, there are two characteristic difficulties here, difficulties acknowledged even by medical practitioners themselves. First, the definition seems insufficiently discriminating since contemporary pluralistic societies are rather tolerant of differences in belief. Though the overwhelming majority of scientists disbelieve in, for example, horoscopes, it is possible to express great faith in astrological predictions and to hold that faith incorrigibly in our society without being regarded as delusional. One can even command a high salary from popular newspapers for consistently espousing these false beliefs, without any fear of attracting the attention of the psychiatric profession. No refinement of the definition has yet been provided which succeeds in making it sufficiently discriminatory. Accordingly, this criterion alone cannot suffice as guidance for psychiatric personnel in their everyday diagnostic work.

Worse still from the point of view of the customary understanding of these matters, the diagnosis of delusional disorders appears to be made without checking off the points in the definition. One of the delusions cited by Palmer concerns a man who claimed to have met Thor in a field in the English midlands. Yet the falsity of such claims is not investigated empirically by, say, field trips to locations thought to be popular with northern-European deities. Accordingly, the official definition of a delusion seems neither adequate for defining the disorder in a precise sense nor to be actually used in identifying whether patients' experiences are delusional or not. Even if the official definition does succeed in capturing a sense of the uneasiness that is felt in the presence of delusional talk, it does not explicate how the haecceity of 'suffering from delusions' is recognised.

Palmer responds to this apparent inadequacy in the definition not by challenging the 'construction' of the definition but by seeing the definition as an impoverished, simplified version of the actual practical skills deployed in recognising delusional accounts. He uses a close-grained analysis of the talk of patients suffering from delusions to try to identify how the delusional properties of that talk are identified. A key finding is that delusional talk differs systematically from the talk of people who, though describing paranormal phenomena (such as sightings of ghosts), are not seen as delusional. In brief, in the accounts of the latter category of people, they attend to the paranormal nature of the events they are describing. They anticipate scepticism in their audience and, in the very way they talk, typically convey the sense that they were as surprised by the paranormal occurrence as their audience is likely to be. They 'incarnately produce' an experience of shared perceptions, of a lifeworld held in common. By contrast, Palmer documents the finding that the accounts of patients with delusions do not attend to such matters. They treat the occurrence of paranormal phenomena as though they were not exceptional. As Palmer summarises this argument:

> people without delusions are concerned with the grounds on which their stories might be doubted and so attempt to undercut those grounds. Doing this involves substantial engagement with the other person and with their interactional concerns. It involves entering into debate with them and arguing your point of view. It is this 'outward'-looking orientation which is absent from [the analysed patient's] talk and, as such, he appears disengaged from interactional concerns which constitute the normal social world. (2000: 673)

It is this unusual orientation and the associated interactional insensitivity, which, according to this analysis, allow the diagnosis to be made without checking the details of the testimony against the formal definition of a delusion. The evidence of the delusion is 'in and as' the peculiarities of the delusional account.

Palmer's argument achieves two things. First, because the psychiatrist's diagnosis of delusionality is made on the basis of discussions with the patient, Palmer is able to use a study of delusional talk to achieve the ethnomethodological aim of specifying the 'just thisness' of recognising delusions. His answer is that the recognition is done not through comparing the patient's conduct with the official definition, but by using more everyday interactional skills to work out that the patient is 'disengaged from [the] interactional concerns which constitute the normal social world'. Second, Palmer is able to use this novel understanding of the difference between psychiatrists' practice and the official definition to undermine constructionist sociological critiques of psychiatry. He observes that it is pointless for constructionists to employ all their energy addressing the definition when the definition is only a rough and ready rationalisation of the tacit skills on which diagnoses are, in practice, made (on definitions and protocols see also Lynch, 2002). In both these senses, his work fully realises the objective specified by Sharrock and Anderson (cited earlier) when they maintain that ethnomethodological study of the sciences 'prefers to look into the ways in which scientists encounter their phenomena, to examine the ways in which they "come upon these" in the course of their investigations, to see how – for example – their activities in a laboratory comprise' the phenomena in question (1991: 74).

ASSESSING THE ETHNOMETHODOLOGICAL RESEARCH PROGRAMME IN SCIENCE STUDIES

The achievements of Palmer's empirical analysis suggest that the ethnomethodological programme might well have precise analytical dividends and offer a distinctive challenge to the constructionist approach to science studies, even if the other studies reviewed were significantly more programmatic and less empirically rich. However, there are two peculiarities to his study which serve to limit optimism on this score. In common with the other ethnomethodological studies outlined, in presenting his study Palmer is keen to dissociate himself from constructionist approaches to the same phenomena. In this case, though, the constructionist approaches from which he distances himself set out to be rival explanations for the psychiatric disorder. They are not sociological studies about conflicts over the nature or classification or treatment of psychiatric phenomena. Rather they are attempts to explain the attribution of mental illness in sociological terms. This is in strong contrast to most 'interest theory' cases or to studies in the Empirical Programme of Relativism where the analyst's focus is on explaining different parties' support for rival explanatory theories in, say, physics or biology. In such cases, the social analyst of science is not trying to replace one scientific account of the phenomenon with her or his own sociological account but to understand why scientists advance and subscribe to competing accounts of the same phenomenon. For this reason, Palmer's critique of constructionist approaches cannot be generalised beyond the case he has selected.

This lack of clarity about their constructionist 'adversaries' appears to be shared by other ethnomethodological commentators also. For example, in his well-known study of Garfinkel and ethnomethodology, Heritage criticises studies 'about' (that is, not 'of') work practices, including science, on the grounds that they 'tend to generate full and detailed descriptions of such matters as the income, social networks and role relations among the participants, but they are largely silent about the matters which make these occupations significant in the first place' (1984: 298). However much this may have been true of the earliest studies in the sociology of science, it is not what proponents of SSK or even ANT view themselves as aiming to do. While Palmer can certainly make a claim about the originality and persuasiveness of the findings derived from his approach to psychiatric diagnosis, he cannot sustain the broader ethnomethodological argument that science studies approaches in this area are illegitimate. An approach to psychiatric diagnosis from the point of view of SSK would not argue that psychiatric disorders are really the result of social processes of marginalisation. Instead a much more likely orientation would be to examine disputes within the psychiatric community over the recognition, categorisation and treatment of disorders. For example, interest theorists or proponents of EPOR would be likely to be interested in the recent negotiations through which premenstrual dysphasic disorder became recognised as an officially listed disorder or the only slightly less recent process through which homosexuality ceased to be listed in official US psychiatric manuals as a form of mental illness, a classification that persisted into the 1960s (Kitcher, 1996: 207). In such a case, there would be arguments within the psychiatric community (and within public forums including the women's movement and the lesbian-bisexual-gay movement respectively) over the justifiability of the classification. The systematic study of the progress of that controversy, or the present-day controversy over the interpretation of obesity (Kitcher, 1996: 209), would be the province of science studies.

The second unusual feature of Palmer's study is that it deals with an area of medical science in which diagnostic work is carried out pretty much exclusively through the medium of talk. He is able to apply the techniques of conversation analysis to identify the characteristic features of interactions in which delusions are discussed and thus to offer an account of the haecceity of 'being deluded'. But very few areas of scientific practice – even ones studied by ethnomethodologists – share this feature; for example Button and Sharrock (1993: 6) are insistent that Garfinkel et al.'s well-known ethnomethodological study of the work of making an astronomical discovery (1981) is not a story about the linguistic construction of a new phenomenon. For this reason, Palmer's work can just as readily be seen not as the first of a series of ethnomethodological rebuffs of SSK but as an exceptional case: a case which exemplifies ethnomethodology's ambitious project without offering any guarantee that this ambition is achievable in more general terms through other ethnomethodological studies. As noted above, there are other reasons also to come to this unpromising evaluation. Not only was there Bjelic and Lynch's pessimistic conclusion (1992: 74) arising from their study of Goethe and Newton, but Lynch also notes that 'It is difficult to tell whether Garfinkel's project failed or succeeded', largely because those who followed Garfinkel's lead most precisely failed to get jobs as mainstream sociologists (Lynch, 1993: 275).[3] Palmer's study powerfully points out the drawbacks of using scientists' definitions, rather than their practice, as the starting point for constructionist analyses. But the more general argument, that constructionist approaches are illegitimate while ethnomethodology is the best way to analyse scientific practice (see Lynch, 1992), has not been successfully made.

SCIENTIFIC DISCOURSE AS THE FOCUS FOR SCIENCE STUDIES

In recent decades, the term discourse analysis has become very widely used within the social sciences and the label has been applied to a confusingly wide array of techniques and approaches. The focus in this chapter will be limited to one particular approach, principally associated with Mulkay and Gilbert; to distinguish this from all other varieties of discourse analysis, it will be termed the analysis of scientific discourse

(ASD). As with ethnomethodological studies of science, the analysis of scientific discourse is primarily an empirical approach which concentrates on the details of various scientific 'texts' including scientists' formal and informal writings but also transcriptions of scientists' talk.

Mulkay launched the ASD programme with arguments of two main sorts. First, he attacked the existing practices of science studies and the sociology of scientific knowledge in methodological terms. At the same time, with Gilbert, he offered examples of what the replacement form of investigation – the analysis of scientific discourse – could offer. This strategy is very similar (tactically and conceptually) to that adopted by ethnomethodological critics, so much so that Heritage initially cited Gilbert and Mulkay as examples of contributors to the ethnomethodological programme of studies of science at work (1984: 303). Where ethnomethodologists argued that science studies missed the specifics, the haecceity, of science, Mulkay accused science studies of 'vassalage', of merely telling stories woven together out of fragments of scientists' own discourse (Mulkay, 1981). And where ethnomethodology offered to explore the 'just thisness' of mathematical proofs or of computer programming, Gilbert and Mulkay offered an understanding of the culture of science through a systematic exploration of the discourses which recurrently figure in scientists' writings and talk (Gilbert and Mulkay, 1984). The best way to assess the ASD programme is to begin with the discourses they identify.

Two Discourses of Science

The central analytic claim made by Gilbert and Mulkay is that there are two principal discourses which feature over and again in scientists' talk and writing. Yet these authors did not set out to study scientific language but to conduct a study of a biochemical controversy over the way in which biological cells store and handle energy. They conducted interviews with the leading protagonists in the field and collected these scientists' papers and other writings. But when Gilbert and Mulkay came to investigate particular substantive themes, such as how scientists chose between theories or why some scientists supported one hypothesis while other equally well-placed biochemists favoured its rival, they found that the most striking sociological phenomena were the consistencies in the ways in which their respondents talked about and framed the issues (Gilbert and Mulkay, 1980). Where there were rival biochemical hypotheses, respondents systematically disagreed about which hypothesis was stronger, which was better supported by the evidence, which was favoured by the better experimentalists and so on. However, advocates of both sides were equivalent in one regard: they recurrently claimed that their own beliefs were directly supported by the evidence while their opponents' views were impaired by psychological or cultural distortions or other contingent difficulties.

Mulkay and Gilbert generated three linked arguments from these research results. The first of these arguments concerned the existence of two discourses or repertoires which are repeatedly found in scientists' accounts of their field. One discourse, which they term the empiricist repertoire, presents scientific belief as following relatively unproblematically from experimental facts. From numerous instances of such accounts, these authors derive five assumptions embedded in the operations of this discourse (Mulkay and Gilbert, 1982a: 173):

1. That the speaker is able to recognise the experimental facts.
2. That correct scientific belief is determined by the facts, which are themselves independent of the theories under consideration.
3. That there cannot be more than one correct theory.
4. That incorrect belief is caused by and invalidated by 'non-cognitive' influences.
5. That correct belief is independent of such influences.

Speakers appear to employ this repertoire when describing and affirming their own beliefs and those of people with whom they agree. By contrast, there is a contingent repertoire according to which scientists' actions are depicted 'as the activities and judgments of specific individuals acting on the basis of their personal inclinations and particular social positions' (Gilbert and Mulkay, 1984: 57). Mulkay and Gilbert (1982a) find a consistent pattern of asymmetrical accounting in the way that scientists speak and write about their field. While scientists' own beliefs are characteristically presented as arising directly from the experimental evidence and are thus treated exclusively in terms of the empiricist repertoire, their intellectual opponents' views are explained away in contingent terms, invoking the distorting effects of psychological and cultural variables. The key analytical finding is that neither of these accounts can be taken literally at face value. Since both 'sides' in the biochemical dispute account for their own beliefs in empiricist terms, one cannot adopt this repertoire without arbitrarily accepting one side's position and denying their opponents. Neither can one straightforwardly adopt the contingent repertoire since speakers appear to use this repertoire rather 'opportunistically'. The accounts of opponents' weaknesses do not seem to be built with great concern for internal consistency; rather they are characterised by 'flexibility' (1982a: 169; for further analysis of error accounts see McKinlay and Potter, 1987).

Having identified and characterised these discourses, Mulkay and Gilbert use them – in the second step of their argument – to criticise sociologists of scientific knowledge. They claim that qualitative analyses of the scientific community tend to derive their evidence (for example about participants' cognitive interests or about the 'interpretative flexibility' surrounding scientists' findings) from the workings of the contingent repertoire. In this sense, science studies has tended to delegate its analytic work: 'the qualitative analyst tends to allow *participants* to do the analysis' (Mulkay, 1981: 168, original emphasis). But this is flawed because:

> one can show that the regularity found in scientists' *statements* about political action is produced by a regularity in the methods by which they construct their accounts of action, rather than by a regularity in the actions themselves. (1981: 168)

Mulkay follows this claim with the accusation that 'the literature of social studies of science is largely derivative from scientists' own literary products and accounting procedures for its versions of scientific action and belief' (1981: 171). This is the basis for his charge of vassalage.

In principle, the remedy offered by the ASD sounds similar to the ethnomethodological recipe, in other words to focus 'not on action as such, but on the methods scientists themselves use to account for, and make sense of, their own and others' actions' (1981: 170). Accordingly, in the third step in their argument, Mulkay and Gilbert use studies of the play of the two repertoires of scientific discourse to explore hitherto overlooked aspects of scientific culture. Among the clearest examples of the way in which paying attention to the features of scientific discourse can yield analytical insights comes from Mulkay and Gilbert's exploration of scientific jokes and humour. Noticing that the discourses, though incompatible, are both widespread and pervasive, Mulkay and Gilbert point out that there is constant potential for the discourses to be juxtaposed and for the incompatibility between them to become troublesome. While a scientist may claim her or his views are grounded only in the experimental evidence, intellectual opponents will account for her or his beliefs differently. Mulkay and Gilbert examine various 'devices' for handling these troubles; one such device appears to be institutionalised forms of humour. Mulkay and Gilbert collected various examples of scientists' humour including circulated text-based jokes, cartoons and 'entertaining' biochemical songs.

As an example, many of the biochemists interviewed knew of, and in some cases had posted up in their offices, a table entitled something like 'Dictionary of Useful Research Phrases'. This offered two columns of text, the one on the left purporting to be what a scientist had written and the right-hand one indicating what had truly been meant. The written claim that 'It has long been known that X' translates as 'I haven't bothered to look up the reference', while 'Three of the samples were chosen for detailed study' allegedly equates to 'The results on the others didn't makes sense and were ignored' (Mulkay and Gilbert, 1982b: 593). This elementary joke appears to be pervasive within the research culture investigated and is thus presumed to be expressive of cultural sensibilities. Analysts of scientific discourse suggest that a leading source of its humour derives precisely from the fact that these columns allow the two registers to be juxtaposed. It 'translates' the empiricist repertoire into the contingent one and thus amusingly exposes the potential for tension between the two accounting repertoires, a potential which is normally managed by the systematically asymmetrical treatment of the views of one's allies and one's opponents.

THE STATUS OF SCIENTIFIC DISCOURSES

As well as illuminating one source of humour in the scientific community, Mulkay and Gilbert treat this study as indicative of more general patterns of sense-making in science, since 'Participants' organization of humorous discourse is simply one aspect of their ability to construct diverse interpretations of their social world' (1982b: 606). In part this claim appears to have been made good since, in a subsequent study, Mulkay went on to explore the language of prize-acceptance speeches in science (1984). Analysing the speeches given at Nobel Prize ceremonies between 1978 and 1981 (the period when one of 'his' biochemists received the award), Mulkay detailed ways in which the recipients of the prizes, the Nobel Laureates, redistributed the praise to colleagues, mentors, the sagacity of the entire scientific community and so on, apparently under the dual imperatives of avoiding excessive self-praise and of honouring the seriousness of the occasion. Here too is an area of scientific activity, rarely addressed by sociological analysts, where one can offer analytical insights by viewing the task facing award-winning scientists as being to 'construct an interpretation of their social world', albeit – in this case – with suitable respectfulness and modesty.

But the programme of the ASD faces two linked problems. The first is methodological. Advocates of the ASD need to demonstrate that one can get no deeper than the discourses; as Gilbert and Mulkay put it early on, the adoption of the discourse-analytic approach implies 'that several longstanding questions in the sociology of science are unanswerable in their customary form' (1980: 293). Second, they also need to show that the two discourses they identify are the principal ones available and that these discourses are, in significant ways, internally coherent. Some light is shed on the viability of these assumptions by a later study of Mulkay's that deals with the public debate in Britain about whether to permit research on human embryos or to make it illegal. Using the records of Parliamentary debates on this matter, Mulkay identified two leading repertoires, respectively the rhetorics of hope and fear (1993). In the briefest terms, the rhetoric of hope develops the idea that science and technology have been repeatedly associated with overcoming disease and other forms of human progress; appropriately regulated research on embryos is therefore likely to be a force for good. Those who give voice to the rhetoric of fear, by contrast, allude to concerns about meddling in nature and to humans over-reaching themselves with disastrous results, as commonly associated with the story of Herr Frankenstein and his monster. Mulkay uses extensive quotes from the official records of the Parliamentary discussions to show that these repertoires were recurrently used and that most speakers framed their contributions in these terms. At the same time, Mulkay is able to show that speakers on both sides of the debate are able to use the rhetorics flexibly. For example, opponents of legalisation deploy the repertoire of hope in relation to other scientific endeavours but they design their rhetorical interventions to exclude research on embryos (1993: 733). In this sense, Mulkay's analysis resembles the earlier claims he made when writing with Gilbert since it appears that all the actors involved were able to call on both discourses. However, in this case the underlying model is of speakers skilfully fashioning their contributions to the debate out of existing cultural resources. Thus Mulkay asserts that aspects of the rhetoric of fear 'were an emotive technique designed to convince listeners' that past fears might be repeated in the near future (1993: 738).

As an analytic procedure, this is very different from the earlier assumptions of the ASD approach. Then Mulkay and Gilbert suggested that actors' interests were not available to sociological analysts since talk of scientists' interests was only constructed out of uses of the contingent repertoire. One could not speak of scientists' beliefs since one only had their discourse to go by; to treat anything other than discourses was to re-subject oneself to vassalage. In the later study, the analyst is able to claim that the adoption of a repertoire is a 'technique' which advances actors' political purposes. Previously, the discourses were the fundamental units of analysis; now it appears that there are actors who are opting for particular rhetorical manoeuvres. In his study of the embryo research debate, it appears that Mulkay is operating rather like the sociologists he previously criticised. Furthermore, just as with the identification of the earlier empiricist and contingent repertoires, the rhetorics of hope and fear are identified simply as empirical generalisations. Mulkay offers no interpretation of where these discourses come from; neither does he offer any systematic examination of the internal consistency of these discourses (something to which Latour had objected in his response to the earlier ASD work: 1984).

CONCLUDING REMARKS

In summary, the ASD programme faces severe difficulties as a procedure for investigating science. Part of ethnomethodology's reason for rejecting mainstream science studies was the feeling that the sociology of scientific knowledge missed the just thisness of science. In some respects, Mulkay and Gilbert's arguments appeared to be in accord with this rejection of science studies. But the ASD approach started out by proclaiming that those central questions were simply unanswerable precisely because one could not peel back the layers of scientific discourse to get at actors' beliefs and deeds. In this sense, the ASD is not (in ethnomethodologists' terms at least) a sociology *of* science at all. It expressly rules out the study of the core themes of the scientific lifeworld and thus cannot address the haecceity of doing science, the ethnomethodological grail.

Both ethnomethodology and the ASD make significant challenges to the customary approaches in science studies. Both claim to find fault with science-studies-as-usual and to offer a substitute programme which avoids the faults. Verdicts on the two approaches, however, differ. It is easier to appreciate the appeal of the ethnomethodological critique. As noted in Chapter 4, even Latour has lately begun to laud the ethnomethodological approach and to describe his 'constructionist' work as affined to it (Latour, 2000: 112).[4] Like the ethnomethodologists, he fears that social constructionism threatens to explain science away; full-blown social constructionism is a position by which, he asserts, one could not be convinced for 'more than three minutes. Well, let's say an hour, to be fair' (1999b: 125; see also Hacking, 1999). Despite the appeal of the critique, the viability of the resulting ethnomethodological research programme as a general approach to the analysis of science appears seriously in doubt. By contrast, the ASD research programme gave rise to several interesting studies of hitherto neglected aspects of scientific culture. But the fundamental criticism (the charge of vassalage) appears so exaggerated that it has even been dropped by Mulkay himself in his later studies; most recently he was effectively no longer a practical exponent of his earlier theoretical position.

---

[1]In earlier ethnomethodological writings the term 'quiddity' was used to perform much the same function as haecceity now serves; it was dropped, possibly for fear of being seen as having essentialist implications (for opinions on this see Lynch, 1993: 283–4).

[2]At the time of writing the most recent publications in this tradition were collected in *The British Journal of Sociology* 53 (2002), the whole of Issue 2.
[3]Lynch modestly omits his own considerable professional success, a success which might well lead one to overturn this assessment.
[4]Lynch extends a slightly less warm welcome stating, for example, that 'A challenge I would raise to Latour ... would be to specify what would be lost, for example, if he were to present his natural language description of the Amazon field project [described in Chapter 4] without attempting to subsume it under abstract models of referential "elements" arranged in chains and of hyper-abstract networks of human and non-human "actants". My null hypothesis is that nothing would be lost, though readers might then wonder about the point of the description' (2001: 230). Quite.

7 Reflection, Explanation and Reflexivity in Science Studies

INTRODUCTION

This chapter is concerned with assessing the various lines of thought in science studies reviewed in the previous chapters, with evaluating what they contribute to sociological understanding of science and how they prepare us for the task of the third part of the book. In part, this assessment will depend on a judgement about how successfully science studies accounts for scientific conduct and helps us make sense of the character of scientific knowledge. But before arriving at that point, there is a prior consideration because some authors in the science studies tradition have made theoretical ideas and explanatory work in science studies the focus of their own analysis – in other words they have promoted reflexivity in science studies. That is not to say that these authors were the first to turn their attention to issues of self-reference in science studies. On the contrary, the topic of reflexivity has all along been viewed as a potentially troublesome issue by a clear majority of authors in the science studies tradition.

SSK's core claim – that knowledge is in some sense relative to the social circumstances of its production – when paired with the demand for symmetrical styles of analysis leads very easily to the suggestion that the sociology of scientific knowledge (even ANT) is itself a product of the various authors' own social milieu. This thought can readily be developed to have apparently unsettling implications for science studies. For example, it has on many occasion been advanced by rationalist critics as a 'knock-down' argument against the whole enterprise of SSK (see Laudan, 1982). Thus, somebody who announces that knowledge is relative to actors' interests and who purports to believe that people accept knowledge-claims which conform with their interests rather than with universal, rational standards of argumentation, is – on the face of it at least – acting in a self-refuting way if they offer rational arguments in favour of relativism. According to this line of reasoning, interest theorists (for example) are acting in bad faith when they engage in argument about interests since, according to their own position, it is interests which trump evidence and argument. Worse still, the fact that interest theorists bother to engage in argument implies that, deep down, they do not really believe in their own SSK theory since their practice belies their supposed theoretical beliefs.

Critics commonly carry on as though they were about to take science-studies authors by surprise with these arguments, as though this were an ambush for which sociologists of knowledge were wholly unprepared. But, as we have seen, Bloor and Collins recognised this potential source of trouble from the very outset. Indeed, Bloor attempted to pre-empt this criticism by adopting reflexivity as one of the four tenets of his Strong Programme, even though he subsequently displayed little interest in following this tenet through in practice. Most other people responded to the potential problem either by banning discussion of it – as, for example, Collins sometimes does (Ashmore, 1989: 115) – or by treating the relativism proposed by the sociology of science as merely methodological. In defence of the banning option, Collins proposed that it was the 'natural attitude' of scientific investigators to be somewhat naïve realists in the course of their work; Collins proposes that this day-to-day outlook should apply to the social scientific investigator as well. Alternatively, analysts who emphasise that they are methodological relativists have argued that they are not committed to ontological relativism. When studying a scientific controversy in which – by definition – no one is sure of the 'correct' outcome, they merely find it empirically fruitful to behave as though they were relativists. In this sense they are not 'really' relativists and, in this way, they avoid the apparent self-contradictory implications of reflexivity.[1]

A few authors, however, have veered away from these defensive strategies and selected a different path: they have sought to make a virtue of the seeming problems of reflexivity. I shall examine these authors first and then return to the issues of reflexivity, impartiality and explanation in the sociology of science.

BEING REFLEXIVE ABOUT THE SOCIOLOGY OF SCIENTIFIC KNOWLEDGE

Proponents of reflexivity have argued that, since authors in science studies are faced with persistent problems of self-reference, there is no prospect of higher ground to which they can ascend to avoid it. Science studiers should instead confront those difficulties in their work and in their texts. This reasoning has led authors who celebrated reflexivity to adopt one or other of two main (though not completely distinct) paths. The first of these is towards the adoption of so-called New Literary Forms, the second leads to an ever more acute engagement with reflexivity itself. In the latter camp, authors such as Woolgar and Ashmore have chosen to tackle reflexivity head on, with knowingly ironic invocations of advance and progress. Rather than view the self-application of SSK as an abyss into which one is best advised not to peer, they have interpreted the abyss as a new frontier for analytical investigation. In this spirit, instead of treating reflexivity as a major 'in-principle' problem for science studies, Ashmore was one of the few authors who decided to pursue the topic empirically (1988; 1989). He used Collins' work on replication in physics (largely the gravitational wave community) as one of his case study areas.

As pointed out in Chapter 2, Collins claimed that conventional accounts of scientific method placed great emphasis on replication as the way in which experimental and observational truths were grounded. However, he sought to show that the status of a second event as a replication was itself subject to negotiation. No two experiments could be literally identical; therefore there was always a measure of interpretative judgement in determining whether a second experiment should count as a 'repeat' of the first. A purported replication might, for example, not be counted by members of the relevant community because the experimenter was deemed to be incompetent or insufficiently skilled. Equally, the supposed replication is bound to differ from the original in myriad small ways and these differences may or may not be consequential in determining whether it is sufficiently close to merit the label of 'replication'. This was the basis for Collins' claim about the experimenter's regress. In his work on the 'life and opinions of a replication claim' Ashmore (1988) examines Collins' study of replication in physics. Collins had made two claims documented by Ashmore (1988: 128). First, he had noted (in line with the commentary given above) that, 'The objectivity of science and its insulation from social and political biases are *supposed to be* ensured by ... above all, the possibility of replication of ... work by independent parties' (emphasis added). But, he had also asserted that 'Ultimately the argument for all these things rests ... above all on independent replication of the findings.' The droll disclosure, of course, is that in the first instance Collins is talking (sceptically) about replication in physics; in the second he is talking (straightforwardly) about other sociological work that replicates his own studies of replication in physics.

What Ashmore did was to carry out a Collins-style study of replication in the sociology of science community. He examined a bunch of SSK studies which purport to be, or are claimed as, replications of Collins' work, interviewed the authors, and deconstructed the studies in the Collins mode. According to his text, this led him to the ironic position that he has both replicated Collins' work (by finding that certain studies get counted as valid replications only through social negotiation) *and* undermined the position of all the other candidate replication studies since their credentials for counting as replications can be questioned in each case. Their status as replications appears merely to be the outcome of social negotiations. Rather than interpret this as a basis for a refutation of Collins' position, Ashmore treats it more as an examination of how issues of reflexivity and self-reference are handled in texts. In other words, his focus transfers from questions of overall logical consistency to matters of textual practice (Yearley, 1981).

In a joint essay with Woolgar, introducing a collection of studies in reflexivity (Woolgar and Ashmore, 1988), the programme of reflexive studies is presented as the 'next step' in the sociology of knowledge. They offer a three-step model where analysts move from realist assumptions about natural science and about their own practice, through the relativising of the treatment of natural science, to the symmetrical treatment of their own and their subjects' knowledge. In line with their reflexive outlook, they draw attention to the fragility of this construct by mentioning at the end of the chapter that they could have included a second table on the construction of 'progress' in tables (1988: 10). All the same, they seem to make two main claims to the advantages of reflexivity: that it completes the move towards symmetry and that, furthermore, in explicitly reflexive forms of analysis, novelty is introduced by 'things that start happening' in the text (1988: 5).

The difficulty with these two claims is that there is rather little detailed evidence to support them. As with Callon and Latour's work (discussed in Chapter 4), one can see the in-principle attraction of moving towards generalised symmetry. But the dividends have been meagre, rather more so even than those of ANT (Collins and Yearley, 1992a: 303–11). For example, the nearest thing to a conclusion in Ashmore's study of Collins (1988: 151) actually arises from a claim made by Collins himself in a letter to Ashmore, quoted in the text (Ashmore 1988: 126 and again on 150). Collins' cited statement is that 'the permeability of replication does not mean that [replicability] is still not the only criterion of what is to count as a natural regularity (or social regularity). It is the only one we have.' Faced with the reflected image from Ashmore's analytical mirror, Collins seems to concede that he cannot escape the 'logic' of replication even though it is much less clear-cut and straightforward than its establishment proponents seem customarily to assume. Ashmore observes that Collins comes close to admitting that his (Collins') work is effectively self-exemplifying: it demonstrates the 'permeability' of replication both by documenting such permeability in physicists' work and by manifesting such refractoriness itself in face of attempts to replicate it. Ashmore maintains that, had Collins been quicker to acknowledge the self-exemplifying character of his work, it would have been more persuasive in the first place. It would also have done away with the need for Ashmore to do his own 'replication study' and the associated interpretative work. Ashmore seems content to let self-exemplification be the analytical outcome and to leave Collins' (revised) conclusion to stand.

The other claimed 'dividend' of reflexivity (the appearance of novel phenomena in the text) is best revealed in reflexivists' second analytical tack, the adoption of New Literary Forms; this too has lately been championed by Ashmore et al. (1995). For this text, he and his co-authors invent a novice author who chronicles her introductory survey of the literature; 'she' then writes:

> the most interesting things I have read are those that foreground their authors' own involvement in the text. And it seems the closer one's *topic* is to one's *method*, the more important it becomes to devise some way of coming to terms with the implications of that similarity. In this argument, writing about writing *has* to be a self-consciously circular process. (1995: 339, original emphasis)

In New Literary Forms, innovative styles of writing are used that are designed to make the textual work of representation apparent to the reader. Studies are written as dialogues, plays or diaries in order to accentuate the 'situatedness' of the writer. Dialogues allow the authorial viewpoint to be challenged by a counter-voice as the text proceeds, while the diary format allows the writer's 'thought' to develop as time supposedly passes. Such textual moves avoid the presentation to the reader of an artfully seamless text. Speaking as himself, Mulkay describes his discovery of the advantages of textual self-awareness in the following terms:

> The phrase 'new literary forms' is better than, say, 'new analytical language', because what was needed at that time was not a new vocabulary for writing about social life, but new ways of organizing our language which would avoid the implicit commitment to an orthodox epistemology that was built into the established textual forms of social science. In an attempt to address the self-referential nature of SSK's central claims and to display the ways in which analysts' claims are moulded by their use of specific textual forms, I began to employ multi-voice texts in which both analytical claims and textual forms could become topics of critical discussion in a natural manner. Texts of this kind made it possible, I found, to replace the unitary, anonymous, socially removed authorial voice of conventional sociology with an interpretative interplay within the text as a result of which the voices involved became socially located and their constructive use of language became available for comment both within the text and beyond. (1991: xvii)[2]

In particular, the author experiences, and does not try to conceal, an inability fully to control the text – the kind of lack of control that counted (above) for Woolgar and Ashmore as 'things starting to happen'.

This interest in textual manoeuvres has allowed these authors to ally themselves with developments in literary theory and with other textual experimenters (see the lists in Woolgar and Ashmore, 1988: 2; and in Ashmore, 1989: 225; see also Mulkay, 1988[3]). But two forms of misgiving have been raised about this route into New Literary Forms. First, some authors have questioned whether these techniques are in practice as liberating as they appear. The additional voices in the text are still under the control of the author so that any subversive effects are not genuinely disruptive (Pinch and Pinch, 1988). Second, the ultimate aim of the NLF procedure is imprecise. The objective is certainly to be clear and open about the basis for the claims one makes in one's own text. But there is a remaining ambiguity since it is not apparent whether the analytic interest is in textual devices which are peculiar to the persuasiveness of specific texts, or whether the devices literary analysts identify are simply present in all texts. If it is the latter, as seems to be most likely, the analytical conclusions of any study by Ashmore or Mulkay or Woolgar have little to do with the persuasiveness of any particular text they have studied. Indeed, viewed in this way, the attention to one's own practice takes the procedure close to ethnomethodology and the focus on the 'incarnate production' of credible scientific writing. Handily enough, Garfinkel and other ethnomethodologists had always used the term reflexivity in their own special sense, to refer to the way in which specific social practices relate to a particular sphere of activity. The ethnomethodological claim is that social practices are designed to be adequate to the purposes in hand and, in that sense, all social action is reflexive since it is monitored against the appropriate contextual standards. As Garfinkel says: 'such practices [actors' situated practices] consist of an endless, on-going, contingent accomplishment; ... they are carried on under the auspices of, and are made to happen as events in, the same ordinary affairs that in organizing they describe' (1967: 1). Social life is always and ever reflexive precisely because society is incarnately produced by knowledgeable subjects.[4]

ETHICAL REFLEXIVITY IN CONSTRUCTIONIST SCIENCE STUDIES

Inventive and arresting though this work on reflexivity is, it has been overwhelmingly cognitive. In other words, the worry about reflexivity has been treated almost wholly as though it had to do only with logical consistency. Equally, the supposed benefits of experimenting with New

Literary Forms have been understood as consisting of increased self-awareness and coherence. But there is another concern about analysing science-studies practice which has a much more normative element. As was made clear in Chapter 5, feminist science studies were typically concerned with considering ways in which the organisation and production of scientific knowledge might lead scientific institutions to generate claimed 'facts' and theoretical interpretations inimical to the presumed interests of women. This was made clearest in Longino's attempted resolution of these themes where she distinguished constitutive from contextual values and proposed that, where constitutive values alone cannot lead to interpretative closure in debates about (say) the facts of human descent, then contextual values may legitimately be called on to arrive at a preferred interpretation.

In such cases the concern is not that symmetry and impartiality lead to the logical impasse of self-reference but that they may have morally reactionary results. In other words, many social scientists have been more disturbed by such moral relativism than by the apparently more radical idea of cognitive relativism. This, it should be noted, is contrary to the prevailing traditions in the contemporary Western world where facts are taken as intersubjectively constant while values are held to be matters of conscience and thus legitimately subject to cultural and social variation.[5] Where this normative issue has been addressed explicitly by advocates of SSK, it has typically been written about in terms of 'ethical neutrality'. The question has been asked: if one is to be a cognitive constructionist (and thus be impartial about truth and falsity), does that mean one has to be an ethical and political constructionist (and thus be impartial to issues of justice, political principle and the attribution of rights) too? In debating this issue, the claim has been made (Scott et al., 1990 and Richards, 1996) that to study a controversy and to examine in great detail the respective sides' arguments is inevitably to involve oneself in the substance of that argument since, typically, the economically or politically weaker party will experience difficulties in deconstructing their opponents' case. Any symmetrical deconstructive work carried out by sociologists of scientific knowledge is likely to be disproportionately helpful to the weaker side. Accordingly, science studies – by its very practice – has normative implications, and the route of ethical neutrality is not genuinely available (see also Pels, 1996). We delude ourselves if we try to be neutral in that sense.

Collins has been one of the few science-studies authors among those reviewed thus far to address himself specifically to these issues in print (1996; though see Pinch, 1993 and Yearley, 1993, both reviews of Richards, 1991). Collins argues that it is the analyst's responsibility, as an analyst, to try to be as neutral as possible though he acknowledges that:

> From time to time researchers are likely to discover themselves liking some of what they see and not liking other things, or feeling that some cause must be supported and another countered. What I have argued is that support for such causes should be based on the merits of the causes .... (1996: 241)

There are two curious things about this quote: first the verbs selected to describe analysts' ethical evaluations and second the notion of 'merits of the causes'. Collins refers to analysts discovering themselves liking something: this seems both to imply that the analyst did not deliberate about the matter much (since they only realised they liked something when they 'discovered' that they were liking it) and also the verb 'to like' connotes a rather subjective sentiment. Analysts' fondness for a cause appears to have rather little to do with reasoning and rather more to do with un-thought-through preferences. It seems that Collins introduces into his very choice of words the individualistic, liberal notion that moral causes are simply matters of preference. Second, it is unclear how Collins can separate the merits of the cause from various matters of fact which the analyst is required to set aside in accordance with the methodological injunction to be cognitively relativistic. It is far from clear how the analyst can both be a methodological relativist and – at the same time – take a view on the merits of the case unless the science-studies analyst has bifurcated personae, taking one view as an analyst and a quite different view as a private citizen.

When Collins comes to offer positive proposals about how the science-studies practitioner should approach evaluative issues, his answers are – at first sight at least – surprising. In response to Richards and others, he offers a pro-Mertonian position, proposing that: 'We know that we prefer a science informed by something like the Mertonian norms' (1996: 232). Furthermore, as noted above in response to Ashmore, he had already suggested that he accepted that replication is still 'the only criterion of what is to count as a natural regularity (or social regularity). It is the only one we have' (Ashmore, 1988: 126). In short, it seems that the challenges of ethical commitment and reflexivity lead Collins to advise SSK practitioners to be scientific (rather as Bloor had encouraged them to be). Perhaps unexpectedly, he fleshes out that advice in terms both of a methodological precept (one with which Popper or even Newton-Smith would be happy) and of the universalistic norms of scientific conduct. Collins justifies his support for neutrality, both of an ethical and cognitive kind, by reference to a notion of ideals. Even if the ideals are forever in practice unattainable, that does not make them mistaken as ideals. He asserts that 'a "scientific" approach is a good one, even in the face of our understanding that science is not what we once thought it to be' (1996: 241).

REFLECTION AND EXPLANATION IN SCIENCE STUDIES

The examination of the topic of reflexivity allows us to make a surprising kind of progress in the assessment of science studies. Alongside Collins' espousal of relativism he attempts to maintain a support for ideals. But it is unclear what his grounds for the identification of these ideals are: it could be a report on the ideals he finds prevalent in the scientific community (more or less as Kuhn did with his listing of cognitive values). Alternatively, it could be something closer to the transcendental deductions offered by Bhaskar: these might be ideals Collins has deduced from reflecting on what science has to be like. Given his normative orientation, it seems likelier to be the latter. In my view, it is in relation to arguments of this sort – rather than in connection with cognitive reflexivity – that the lack of self-reflection in the science-studies community appears most striking. In particular, it seems to me that there are three ways in particular that science studies, and SSK specifically, have reflected insufficiently on their own practices – three ways which turn out to be important for analysing the issue of sociology's 'missing masses'.

The first consideration is that science studies has been insufficiently reflexive about what it takes 'science' to be. Rather like the philosophical analyses of science with which it first had to fight for its right to analyse scientific knowledge, science studies has tended to look for what is common to all science. In practice, ANT's story of a sociology of translation, Longino's focus on constitutive values, EPOR's emphasis on the experimenter's regress and ethnomethodology's consideration of the haecceity of science have all implicitly assumed that science is science. This has been tantamount to a form of essentialism. While there clearly are matters of analytical interest which are highly persistent across the types of science and through the historical development of science (for an insightful analysis of the idea of 'types' of science see Whitley, 1984), for a sociological audience there are also questions about changes in the societal role of scientific expertise and in the implicit 'deal' that the scientific establishment has struck with its societal patrons. As discussed in the Introduction, perhaps the leading achievement of the scientific profession has been to win spectacularly large financial support from the state and taxpayers with relatively little accountability. The earliest, obviously modern scientific work was, for all practical purposes, self-funded and justified in terms of the understanding of the natural world which it could generate; its practitioners boasted of the potential for control over aspects of the natural world which it could offer, in principle at least. Subsequently, the contract became more explicitly about a trade of society's financial support in return for diffuse and unpredictable, though occasionally far-reaching, contributions to the productive and military capacity of modern economies and societies. By the opening years of the twenty-first century the role of much state-sponsored science had turned to issues of knowledge and understanding required for regulation and testing, for example in relation to international attempts to fund scientists to work out the likelihood and impacts of climate change, or state-level studies into the potential incidence of Mad Cow Disease (BSE) and related disorders (Yearley, 1997).

To some extent, this point has already been picked up in the broader science policy literature. Gibbons and colleagues (1994; see also Nowotny et al., 2001) proposed that recent decades had witnessed a changeover from what they termed Mode 1 to Mode 2 science. By this they mean that, increasingly, novel knowledge-claims are being made directly in the context of application by teams of experts drawn from a range of disciplinary and institutional backgrounds. Examples illustrating their point of view would include new work in computer science or in aspects of the human genome; in such cases the new 'pure' knowledge is simultaneously applied understanding. Innovative computing processors are both new knowledge and new products. Commercial and practical considerations are immediately intermixed with more conventional epistemological considerations about the quality of new science. However, their analysis tends to treat Mode 2 as a coherent framework in which all the elements arise together. Though the Modes may in some sense operate as ideal types of the way in which knowledge production could be organised, there is no overwhelming reason why all of the putative elements of Mode 2 should operate together. Whether they are correct in detail or not, these authors effectively make the point that the mode of scientific production is not fixed for all time. Science in the twenty-first century may be systematically different from its nineteenth-century precursors. Other sociologists had noticed similar developments: in one sense one could paraphrase Beck's well-known argument about the 'risk society' as a claim about the crisis of regulatory expertise (Beck, 1992; see also Chapters 9 and 12). The social role of scientific experts is more and more to pronounce on the safety of our technologies; their regulatory ability is under closer and closer scrutiny. Science-studies writers' lack of reflection on the social role of scientific knowledge has left the sociology of science isolated from other sociologists' concerns with science and technology. At the same time, proponents of the sociology of scientific knowledge have failed to convince the broader sociological profession of the value of their distinctive analyses in relation to these current issues surrounding expertise in public.

As well as being insufficiently reflective about the timelessness of science, many analysts within the science-studies tradition have been unreflexive about their own explanatory practices. Indeed, it has sometimes seemed that methodological self-consciousness in science studies can only ever take the form of the reflexive move exemplified in Ashmore's work. The versions of science studies which have been most expressly theoretical, notably the Edinburgh School and ANT, have found themselves limited by their own theoretical apparatuses. Proponents of both approaches have moved away from their rather formulaic (though more theoretically straightforward) beginnings. Science-studies writers have paid rather little attention to broader trends in sociological work, often because their focus has been very much on the epistemological territories over which the earliest battles with philosophers were fought. By the time that Collins rediscovered the attraction of Mertonian norms, the sociological question about how exactly norms and values can be used to explain people's (including scientists') conduct had long been excised from the thinking of most sociologists of science. Collins' own EPOR studies, marked by very acute insights into issues such as the 'experimenter's regress', were none the less characterised by a rather descriptive orientation and an adamantly straightforward empirical approach to their subject matter.[6]

There is a third sense in which self-reflection has been lacking. SSK, ANT and feminist science studies have had little to say about the sociology of the internal organisation of science. Ironically, the best-known terminology for describing intellectual and social structures within science still derives from philosophical sources, from Kuhn's claims about normal and revolutionary periods of scientific development or from Lakatos' proposals about core commitments at the heart of a research programme surrounded by a protective belt of more readily changed beliefs. Despite a wealth of case studies and elaborate arguments about how best to practise science studies, SSK has produced only a small vocabulary for describing the structure of scientific beliefs or the structure of the scientific community. Admittedly, early work in ANT offered a large range of terms with broad apparent applicability (obligatory passage points, interessement, black-boxing and so on).[7] But these were progressively dropped by Latour himself and proved to be of limited explanatory value even if they were descriptively rich. Aside from Whitley's work (cited above), science studiers even paid little attention to the analysis of systematic disciplinary differences among the sciences.

TAKING STOCK

Without going back on my earlier arguments and writing in a New Literary Form, it is none the less reasonable to offer two alternative conclusions to this chapter and to the second part of this book. According to one, science studies has not been terribly successful in its averred aims. The interest theory of the Edinburgh School had an elaborate theoretical orientation but was unable to achieve the precise explanatory successes it sought. Actor-Network Theory achieved considerable success in terms of its apparent adoption by practitioners of science studies and historians of science, but achieved this popularity largely by virtue of offering a flawed compromise that appeared to allow everything to have equal explanatory potential while actually 'explaining' things only in wholly reactionary (in an epistemological sense) terms. This extremely moderate success was acknowledged by Latour in his progressive moves towards a more ethnomethodological, 'constructionist' position which, as Lynch noted, appears to run simply on a parallel track to everyday epistemologies. Ethnomethodology has a strong (and frequently repeated) problematic but few worked-out examples of what its analytical achievements might amount to; moreover, in its ambition not to explain away (to 'ironise', Woolgar, 1983; see also Garfinkel, 1996: 6) the achievements of everyday action it actually leaves science wholly undisturbed. It espouses an indifference to the practical day-to-day achievements of the scientists it studies. Feminist science studies was successful in terms of its feminist empiricism which promoted scientific reform from within the natural sciences, but attempts to develop a more comprehensive critical position largely fell foul of problems with essentialism. The most elaborate alternative, that of Longino, effectively ceded the majority of ground to realism. She took the constitutive values of science as well established and only in need of questioning when they were insufficient to make sense of scientific findings, whereupon contingent values could be invoked. Reflexivity seemed not to bear huge dividends, however entertaining its products, and even EPOR appeared to run into difficulties with its ideal of neutrality, so much so that its major proponent ended up invoking Mertonian norms and proclaiming the value of replication. It took us full circle to re-discover what we thought it had undermined. In a sense, Collins too appears to favour leaving science unmolested.

But the alternative telling finds valuable conclusions by slicing the phenomenon in a different way. The success of science studies lies not in the triumph of a particular school but in more diverse achievements. The first achievement is Bloor's and Collins' 'finding' about 'finitism'. Science studies insists that ultimately it is people or communities who decide on how the world is. Of course, they typically do this based on as much evidence as they can generate. But, in the end, people decide; the world does not. This leads to the second conclusion, that people are critically dependent on each other for determining what is known. The worth of knowledge is decided in communities. Accordingly, people are reliant on each other for guaranteeing the quality of their beliefs, and the kinds of quality-control institutions and procedures they invent are critical. Third, as Shapin has famously argued (1994), relations of trust in those communities are central to the ways in which the value of knowledge is established and maintained. Scientific communities depend far more on trust than standard philosophical models imply. Without an underpinning of trust, claims to knowledge can be gnawed away and undermined. Finally, the production of knowledge demands judgement. Scientists need to decide which experimental results to discount as unreliable, whose claims to pay attention to, and so on; this is just the opposite of Popper's seemingly reassuring idea that the worth of science is secured by a standard – almost mechanical – method. As will be seen in the studies in Part III of this book, it is these rather more modest conclusions from science studies that allow us to re-analyse science in society and to reconsider sociology's missing masses. I shall start this re-analysis by examining the performance of scientific expertise in public contexts.

---

[1]If anyone were still not satisfied with these defences, it would finally be open to science studiers to argue that even if there is a problem about reflexivity, scientific realists have no business attacking sociologists of science on this score since even conventional philosophers of science admit that scientists themselves sometimes tolerate anomalies in their scientific practice: this is just another anomaly which will have to be endured until a resolution is found.

[2]After citing this very clear quote from Mulkay, I realised that Lynch (1993: 107) had cited it too. I take this replication as testimony to the perspicuity of Mulkay's text.

[3]For example, in his reflection on the theme of replication, Mulkay uses a parallel with a short story by the avant-garde Argentinian writer Borges. In Borges' story, a fictional turn-of-the-twentieth-century author undertook as his artistic endeavour to re-compose *Don Quixote* (originally written in the early 1600s), word for word. In this way, two apparently identical texts were produced. But the perverse thing is that (in the story) the later one is not treated as a mere copy but as a superior literary creation. Like Cervantes, the fictional author writes of '"Truth, whose mother is history, rival of time, depository of deeds ... ." Written in the seventeenth century ... this is a mere rhetorical praise of history ... [Authored by] a contemporary of William James ... the idea is astounding' (cited in Mulkay, 1988: 90). Borges' story hints that, in literary composition too, the significance of replications is also a matter of skilled accomplishment. Mulkay views the textual handling of replication in science as but one more example of textual treatments of sameness and difference.

[4]For related reasons Latour too is critical of the pursuit of Ashmore-style reflexivity, which he terms 'meta-reflexivity' and describes as a 'suicidal attitude' (1988b: 166 and 169). He compares the reflexivists' tricks to those of the surrealists, finding

the reflexivists the poorer for having come so much later. Having already asserted that the realist/constructionist dispute takes place in a flat world, Latour sees no prospect for rising above that plane by simply applying that debate to itself. Here again, Latour claims distinctive benefits for his own programme, conferring on his programme the label of infra-reflexivity, claiming that it asks for no privileges but places itself on the same level as the texts with which it deals. Thus, 'When I portray scientific literature as in risk of not being believed and as bracing itself against such an outcome by mustering all possible allies at hand, I do not require for this account any more than this very process: my own text is in your hands and lives or dies through what you will do to it' (1988b: 171). Since one cannot explain the natural in terms of the social, the social scientist faces the same struggles to be believed as the natural scientist or the technologist; that is reflexivity enough for Latour.

[5]This is perhaps out of tune with the ethos of 'political correctness' which prevails in many academic cultures. This tolerance for counter-commonsensical cognitive relativism, particularly if accompanied by intolerance towards moral relativism, accounts – it seems to me – for the extremely frosty reception of aspects of science studies in influential sections of the natural scientific community; for evidence of this see Gross and Levitt's *Higher Superstition* (1994).

[6]This no-nonsense empiricism also characterises the Golem books (see Collins and Pinch, 1993) which are virtually free of all sociological jargon.

[7]It is important to acknowledge other terms of art from work in science studies, albeit not directly from the 'schools' reviewed here, that have been widely adopted. An illustrative list would include the (already mentioned) 'boundary work' (Gieryn, 1999) and 'boundary objects' (Star and Griesemer, 1989).

PART III: SCIENCE STUDIES AT WORK

8 Experts in Public: Publics' Relationships to Scientific Authority

INTRODUCTION: THE PUBLIC'S PROBLEMS WITH SCIENCE

In many advanced industrial countries, but perhaps above all in Britain, the close of the twentieth century and the early years of the twenty-first have witnessed repeated conflicts over the public's trust in and acceptance of scientific expertise. Without trying to be comprehensive, one can list several outstanding issues about which there was evident public disquiet despite attempts by the scientific establishment to suggest that there was little reason for anxiety. In many respects, the clearest case was about the MMR vaccine, a combined vaccine given to young children to inoculate them against measles, mumps and rubella. Isolated studies and some parents' testimonies suggested that there was possibly a rare connection, the exact details of which were not well understood, between the vaccine and children's susceptibility to bowel disorders and (through subsequent transmission of the harmful agents to the brain) to autism. During 2001 and 2002, most of the medical establishment agreed that there was little reason to believe the dissenting voices, but the public was not so easily reassured. In part this appeared to be because the possibility of autism was so terrifying that parents thought there was little benefit in taking a chance, however unlikely it was that the injection would precipitate the condition. In part, there was a suspicion that the MMR injections were being insisted on for reasons of economy; MMR reduced the need for vaccinations by a factor of three and thus fitted with the government's attempt to rationalise and harmonise healthcare. Many parents and other commentators – in newspapers for example – were suspicious of the official policy and of the mainstream scientists' reassurances. No matter how adamant scientists were that there was no cause for concern, public scepticism seemed to persist.

Similar issues had arisen around three years before about the safety of genetically modified plants and foodstuffs. Here again, the government's advisers, institutions within the scientific community and leading regulatory agencies had taken the view that the new foodstuffs were essentially identical to existing crops and that there was no significant danger to consumers at all. Establishment scientists were less sanguine about the potential environmental impacts of the new crops; there were some conceivable problems concerning the impacts of changing farm-management practices on wildlife and about the spread of introduced genes, both of which might merit further investigation. But again there were occasional dissenting voices suggesting that there might just be direct dangers to consumers from GM food. Shoppers appeared to take these concerns seriously and all the leading UK supermarkets competed with each other to withdraw GM foodstuffs from their shelves. These two leading concerns follow hard on the heels of worries, raised a decade earlier, with respect to 'Mad Cow Disease' (more properly, bovine spongiform encephalopathy or BSE). Here again, the government's scientific advisers, over several years, repeatedly advised that there was no evidence of potential harm to consumers. In the mid-1990s there was a sudden turnabout in official opinion and it was officially stated that the cattle disease could be transmitted to humans who ate infected meat. New regulations about which sorts of beef could be eaten were speedily introduced and the European Union imposed a ban on beef exports from the UK. The British livestock industry was also further affected by the long outbreak of Foot and Mouth Disease in 2001 where, once again, scientific advice about how to control the spread of the disease was widely viewed with suspicion. Vast numbers of animals were killed (not all in the most humane of ways) and many of their bodies were burned on huge pyres, causing both moral disquiet and pollution to neighbouring communities. Pollution problems were especially pronounced when the pyres were fuelled with old railway sleepers (railway ties) themselves laced with environmentally troubling wood-preservatives.

These well-known cases raise a number of important issues, not least about scientific advisers to government and about citizens' perception of risk. Thus, in part, the MMR case depends on parents' interpretation of the risks their children face while the 'Mad Cow' example raises questions about how government ministries selected and interpreted their expert advisers. These topics will be treated explicitly in Chapters 9 and 11. My interest at this point is in making sociological sense of public responses to expert claims. On a conventional view of scientific knowledge, there would be relatively little to say about the public understanding of science. The scientific community knows best and most attention would focus on the factors shaping the public's attitude to the scientific community's authority. In the rest of this chapter the analytic claims introduced in the first two parts of the book will be applied to this issue, resulting in rather contrasting conclusions about the public's relationship to expertise.

ASSESSING AND MEASURING THE PUBLIC'S UNDERSTANDING OF SCIENCE

As though in anticipation of the problems with MMR and GMOs, academic interest in the public understanding of science (PUS) arose in its most recent form in the UK during the 1980s in a particular economic and political context. The scientific establishment was feeling squeezed from two sides. Despite the fact that Prime Minister Thatcher was herself a scientist by training, successive government administrations appeared less interested in science and technology than the scientific community felt was appropriate. There was little governmental interest in any science which was of no demonstrable commercial potential. At the same time, the government felt that the most commercially useful research should not be paid for out of the public purse. While Conservative governments had no ideological antipathy to natural science, they tended to take the view that industries knew best what they needed in the way of research and should accordingly pay for it themselves. Even though scientists and engineers argued strenuously that the costs of research were rising quickly, politicians saw no compelling reason to increase public generosity towards the sciences.

Alarmed by dwindling political support and the feeling that accountants were taking over the laboratory, the scientific community also found the public uninterested and, if anything, rather inclined to question scientific advice on matters affecting the public or civil society. Annoyed that newspapers would run astrology features in preference to science stories, leading members of the scientific community felt that innovative concerted action was needed to reassert the importance of scientific knowledge and of the scientific method in public life. Worst of all from the viewpoint of the scientific community, the trends in public and governmental attitudes threatened to be mutually reinforcing. If the public would not speak up for the scientific community it was harder for scientists to get politicians to grant them appropriate respect or to increase their research funds. And if even the government did not seem to treat science as a priority, what lesson was the public supposed to derive from that? From these circumstances the topic of PUS emerged.

At the same time, European and North American industrialists were concerned with the public acceptance of technology (PAT), a theme which, in essence, revolved around a similar set of issues. Commercial representatives insisted that they had new technologies which would benefit consumers and the economy, whether in the health, agricultural, computing or energy sectors. Yet they considered that people were resisting these innovations, usually on unreasonable or spurious grounds. Given the closeness of scientific research and technological innovation, particularly in emerging areas such as biotechnology which could readily be seen as examples of 'Mode 2' research (see Chapter 7), there was in practice considerable overlap between the concerns of PUS and of PAT.

In this context it was felt that key issues which sociological research might address concerned: (i) the extent of public knowledge/ignorance of science and technology; (ii) the most effective ways of communicating with the public about scientific and technical matters; and (iii) the ways in which members of the public thought about science and about those matters on which science and technology had most bearing. Even when approached in these apparently straightforward terms, the social scientific measurement of the public's understanding of science is by no means easy. For one thing, there are different components to the public's understanding of science. In part, PUS is a question of what the public knows about the *substance of particular scientific propositions*. One may wish to gauge how much people understand science by finding out what percentage of the public understands what an atom is, or the differences between bacteria and viruses, or the nature of the greenhouse effect. Naturally enough, it is difficult and expensive to test large numbers of people on a wide range of questions. Alternatively, one may wish to find out something about what is understood about the *nature of science*, where this is broadly interpreted as a philosophical, methodological issue. Do people know why scientists reject astrology or which practices (such as replication) scientists take to be central to the scientific method? Finally, people's understanding of science is likely to be tied to their interest in it, so that attempts to measure PUS have also examined the public's *attitude towards science*.

In fact, before the 1980s most social scientific studies of PUS had focused only on people's attitude to science. Subsequently it was suggested that it might be possible to ask people 'quiz questions' about science, thus throwing light on public knowledge of the content of science. In pilot studies it turned out that respondents seemed not to mind answering these questions when asked in an easy-going format.

The results of these first surveys of scientific understanding among the British public achieved wide publicity (see Durant et al., 1989). On the face of it, the study appeared to reveal shockingly low levels of knowledge, even of the most routine parts of science. One celebrated piece of headline news was that only something around one in three people seemed to know that the earth travels round the sun and that it takes one year to complete its orbit. Those seeking consolation were better advised to look at the mean score. This was rather better than this shock news might lead one to suspect: 11½ correct answers out of a possible 20. It should be noted, however, that some of the questions were markedly easier than others so that there was little surprise that a few questions were answered overwhelmingly correctly. Also, given the admitted variation in the difficulty of questions, the allocation of equal 'marks' for each correct answer is potentially misleading. One conspicuously positive finding was that respondents scored well on a question about probability relating to the likelihood of children inheriting a genetic disorder, despite a widespread assumption that lay people are unfamiliar with the principles of chance.

The same survey included questions about the public's interest in science and about their estimation of the standing of different disciplines. Durant et al. observed that:

> For virtually everyone, medical research is the most interesting branch of science; but for those whose acquaintance with matters scientific is fairly slight, it seems to occupy a truly dominant position. It is judged to be not merely far more interesting, but also far more scientific than anything else. For many of our respondents, it is as if science as a whole were being perceived in terms of what was known about medical science. (1992: 171)

In other words, on average medicine is viewed as the most scientific as well as the most interesting discipline. Survey work conducted in France appeared to confirm this exemplary status for medical research. Durant and his co-authors proposed that this may have important implications for the public's overall perception of science. For one thing, medical research is – in principle at least – clearly aimed at the public good. Medical science without at least a background ideal of healing the sick makes no sense. In this respect it is unlike astronomy or botany, which are more nearly ethically neutral, and other forms of research (such as research on nuclear energy) which may sometimes be seen as antithetical to the public good. In this way, it appears likely that the predominant position of medicine skews public attitudes to science in general towards a more favourable view than would otherwise be likely. At the same time, however, the fact that medicine is an applied form of knowledge tends to foster a utilitarian view of science in the public mind. According to Durant et al.'s interpretation of the survey data, the clear majority of people take a utilitarian view of science and it is the minority who are most knowledgeable about science who tend to place a high value on the extension of basic knowledge for its own sake.

Despite the interest which the results of these surveys aroused, there are unavoidable limitations. For example, given the constraints on the number of quiz questions it is possible to ask (23 in the initial UK research, only 12 in a European follow-up study) and the fact that harder questions were left out because they were so unsuccessfully answered as to be statistically uninformative, it is not possible to use the survey to find out much about what the public does know. In any case the survey is administered as a quiz; the questions are asked out of context. People who do well in pub quizzes are not necessarily the best able to apply the knowledge they demonstrate in their local hostelry. The implications of some of these limitations in the method are revealed by the fact that when asked in a US survey about the characteristics of the scientific method only around 14 per cent of respondents mentioned the key role of experimentation. However, when a similar question was asked using a prompt, the proportion identifying experiment as central to scientific procedures rose to around 56 per cent (Wynne, 1995: 367). It is not therefore easy to say whether 'a mere 14 per cent' or 'well over half' of the public appreciate the significance of experimentation. The answer critically depends on how the question is asked.

Even accepting these limitations on the survey approach to assessing the public's understanding of science, there are still useful conclusions to be drawn from this work. For one thing, it is interesting to note that, even using these kinds of measures, one can see that public acceptance of scientific innovations and optimistic attitudes towards science do not automatically relate to people's knowledge of science. In other words, the most scientifically 'literate' sections of society and the most scientifically 'literate' nations are not the most deferential to science (see Evans and Durant, 1995). Contrary to the simplistic notions of PAT, just encouraging the public to become more knowledgeable about science will not make them more automatically accepting of scientific authority. It seems that knowledgeable citizens may become discriminating 'consumers' of scientific expertise.

PUBLIC IGNORANCE OF SCIENCE OR PUBLIC DISAGREEMENT WITH SCIENTISTS' BELIEFS?

Further consideration of the key question about the link between public understanding of science and deference to scientific authority quickly reveals one of the key problems with the survey method's framing of the issue. The quiz format necessarily assumes that scientists are in the right and that all that is in question is the extent to which the public has 'got' the right answer represented by the scientific belief. This issue has been at the heart of considerable controversy within the social scientific community concerned with PUS.

Critics of the survey approach have argued that the whole focus of the methodology is on how far short the public falls. Such an interest has accordingly been dubbed the 'deficit model' of the public's understanding of science. Many commentators have wanted to reject this deficit model because it plays down the extent to which publicly important science may be uncertain or ambiguous (for example establishment views about the safety of nuclear installations or official advice about diet), and because it implies that all public disagreements with science are due to ignorance.

If we look at some of the quiz questions we can see how dubious these assumptions may be. For example, one of the surveys asked whether it was true or false that 'Natural vitamins are better for you than laboratory made ones.' The clear majority of people in the UK (nearly 70 per cent) got this wrong by answering that it was 'true'. Chemical science indicates that any molecule of the vitamin will be practically identical irrespective of how it was formed. The public's error here can be instructively compared with the response to the question which asked whether electrons are smaller than atoms, a question which only 31 per cent of respondents got right. It seems unlikely that people have an alternative theory of atomic structure according to which electrons are somehow bigger than atoms. Indeed, the lack of public commitment to answers to this question is probably quite reasonably reflected in the over 45 per cent of people who answered that they did not know which was larger. By contrast, it is possible that many respondents who got the vitamin questions wrong may actively believe that natural vitamins are distinctive and possibly healthier. In this sense it might be more accurate to say that they disagree with or are doubtful of the scientific view than that they are ignorant of it. A similar point can be made about the 41 per cent of US respondents who answered that it was false that 'human beings as we know them today developed from earlier species of animals'. This number of respondents is likely to contain many people who are fully aware of what scientists think; they simply reject it.

This is not in any way to imply support for anti-Darwinian views or for 'New Age' views of vitamins. But it is to point out that there is a sociologically important distinction between those bits of science which people simply do not know and those bits which they choose (with whatever prejudice or whatever good reason) to deny. Particularly in the environmental and health arenas, members of the public may argue for example that scientific advice has been wrong in the past (early assumptions about the safety of pesticides, the harmlessness of CFCs and so on have been retracted) and that in future 'artificial' vitamins may turn out to be biologically distinguishable in some way from natural ones.

Advocates of the questionnaire and quiz studies argue that there is still something to be said for examining the deficiencies in people's knowledge. They contend that there are many areas of science about which scientists are strongly agreed and where, notwithstanding arcane points about the provisional character of all scientific knowledge, scientists are not likely to revise their opinions. They claim that it is a question of public importance to find out whether or not the public is ignorant of such matters and to investigate ways of encouraging the public to become better informed.

As this argument is in danger of becoming 'academic', the key question is in what ways (if any) it matters whether or not lay people have the sort of knowledge about science which is likely to be reflected in successful answering of quiz questions. Other research on a related theme suggests that people are more active in their reception and sifting of knowledge than the PUS survey work would tend to imply. A graphic example comes from differences between Catholics and Protestants in Northern Ireland over beliefs about the safety of nuclear power, as indicated in responses to a British Social Attitudes survey (see Yearley, 1995). Apart from socio-economic class I members (who appear comparatively sanguine about nuclear power whatever their background), between 52 and 65 per cent of all other classes regard it as extremely or very dangerous to the environment. Men (60 per cent) and women (61 per cent) are even more like-minded. The variables which are more polarised are the religio-political ones. Thus, 72 per cent of Catholics, 64 per cent of 'Others' (that is, neither professed Catholics nor Protestants) and 53 per cent of Protestants class it as dangerous. On closer examination this appears to be less a matter of religious faith than a political or ethnic issue since, if one looks at percentages regarding nuclear power as dangerous by respondent's professed national identity, we see the figures as set out in Table 8.1 (Yearley, 1995: 132). In other words, ranging the available 'identity' labels from the most pro-British to the least, views on this apparently technical matter of nuclear safety appear approximately to vary according to the 'British-ness' of the label.

On the even starker divide arising from asking people whether, in a dispute, they tend to take the side of the British or of the Irish government, the respective percentages are 54 and 80. It seems likely that the additional antipathy expressed by nationalists is fuelled by a distrust of the state which governs and regulates the nuclear industry.[1] But the main conclusion in relation to PUS is that the best indicator of whether the citizens of Northern Ireland regarded nuclear power as hazardous was their political orientation, not their gender, not their social class position, not even the amount of education they had received.

TRUST IN THE PUBLIC'S RELATIONSHIP TO SCIENCE

The survey quiz questions ask about context-free science. But in everyday situations people have to use scientific information in a context-sensitive way. And trust is central to the contextual assessment of scientific knowledge.

Table 8.1: National identity and the perceived dangers of nuclear power

RESPONDENTS' SELF-IDENTITY[2]	British	Ulster	Northern Irish	Irish
Percentage viewing nuclear power as dangerous	55%	63%	59%	77%

The issue of overriding importance is that people do not experience scientific expertise in a pure context, freed from imputed interests and other background expectations. It is people's typical experience that they receive scientific information for a purpose, for example to persuade them that one washing powder is better than another, that meat is (or is not) a core ingredient of a rounded diet or that nuclear power is a safe and dependable component of a national energy strategy. Since expertise is so commonly related to the experts' (or the experts' bosses') practical agenda, people evaluate the information in the light of their regard for the organisation disseminating it and of any ulterior purpose which they believe they can spot. Furthermore, since the scientific advice surrounding dietary, environmental or other public controversies has itself so often been polarised, scientific knowledge in – so to speak – its natural condition, is increasingly seen as being politicised and not disinterested.

Though pure, disinterested inquiry has always been held up as the exemplary model for scientific inquiry, the science with which members of the public have routinely to deal is as likely to come from the commercial sector. Accordingly, the apparently benign idea that scientific understanding can be smoothly diffused to the public and that the public can readily accept the claims of scientists because they are based on disinterested investigation, can very easily become inverted to look like a charter for pulling the wool over the public's eyes. If not all scientific claims are disinterested, then there is clearly a danger that the 'diffusion' of science will end up lending legitimacy to some questionable claims.

These reservations about the mainstream emphasis in the public understanding of science certainly have not been ignored. They have probably been taken most seriously in relation to public concerns over risk (see also the next chapter). The assessment of risks, whether of new technologies or from natural hazards, has long been an issue in science communication. In the field of risk communication it is clear that the public has a strong interest in weeding out partial and tendentious claims. But once it is acknowledged that trust in the source of information about a risk or hazard is important to understanding people's response, the typical move among policy-makers and students of policy analysis has been to try to break the phenomenon of 'trust' down into its various components.

For example, psychologists may attempt to model the procedures people appear to use in determining the trustworthiness of various sources, concerning for example the perceived expertise and public interest orientation of the body. Or they may look at the credibility which the public attach to different types of organisation or different communication media. The guiding assumption is that the resulting information can then be used to make scientific organisations more credible and to diminish the likelihood of (irrational) non-acceptance of their pronouncements. In an exactly analogous way, where arguments about trust and credibility have been taken on board in relation to the public understanding of science, the typical response in the scientific community has been to supplement an interest in the public's understanding of science with a study of factors affecting public trust in science. To put it another way, confident in the correctness of their scientific views, science communicators see public distrust as a distortion, a problem to be overcome; they aim to find approaches which prevent science's signal being disrupted by the noise of distrust.

There are, though, two shortcomings with this response, both as a practical and an intellectual matter. First, it tends to imply that trust is only an issue when one is dealing with non-specialised audiences for expert knowledge. But this is to fail to notice that trust is central to the business of science itself. Scientific research is not conducted by automata but by participants in a scientific community. Indeed, as noted at the close of Chapter 7, 'Edinburgh School' analyst of science Shapin has recently made the case that the founding of key institutions of modern science in seventeenth- and eighteenth-century England was crucially dependent on newly consolidated conventions of trust and civility (1994: 65–125). Despite the elaborate impersonal mechanisms of peer review and the formal methods learnt by scientists during their long training, the scientific life turns on trust. In particular, scientists cannot independently check every detail of every claim made, whether in routine 'normal' science or in a controversy. As discussed in Chapters 2 and 3, at the forefront of the creation of new knowledge no one may fully know what factors are going to be influential: measurements are often being made at the outer limits of sensitivity of apparatus or at the very edges of computing capacity so that minor differences in the configuration of equipment may have disproportionately large implications. Under normal conditions all of these things are taken on trust. But in a controversy each of these trusted issues can be opened to doubt: long-held assumptions get called into question; the trustworthiness of other scientists and even of the peer review system itself can come to be doubted. In the case of present-day gravity-wave physicists reviewed at the end of Chapter 2, it will be recalled that scientists in one research group went so far as deliberately to remove time and date identifiers from data before passing it to their supposed collaborators because they wanted to remain in control of announcements of coincidences between the two teams' data. Once trust is called into question, the possibilities for distrust begin to expand exponentially.

Given that judgements about trustworthiness are so central to the practice of science, there is little hope that they can be eliminated from the processes by which the public comes to acquire and assess scientific knowledge claims. Trust is an indispensable component in the creation and passing on of scientific knowledge; it is not a feature restricted to lay audiences for science which can be technically manipulated to promote 'high trust' conditions.

The second point is that trust and credibility are not fixed dispositions, either of individuals or institutions. They are the outcome of interactions and negotiations. This is a key point which emerges from several recent qualitative studies on PUS including Wynne's well-known investigation of Cumbrian sheep farmers' responses to scientific advice given in the aftermath of the 1986 Chernobyl fall-out (Wynne, 1995). As a result of radioactive contamination detected after the cloud of fall-out from the Ukraine passed over Britain, far-reaching, though temporary, restrictions were placed on the sale of livestock. Initial confidence that the official scientists understood the problem gave way to public scepticism as the quarantine period was progressively lengthened. According to Wynne's analysis, when farmers encountered the messiness of day-to-day science, when they saw how radiation readings could vary over small distances, or how difficult it was to get a stable figure for the background radiation, the farmers revised their notion of scientific knowledge. This change is neatly captured in a story he relates concerning the live monitoring of sheep, where a farmer saw that out of a sample of a few hundred sheep, just over ten failed the test. They were too highly contaminated for release. Then the farmer recounted how the monitoring scientist 'said, "now we'll do them again" – and we got them down to three!' (Wynne, 1992a: 293). Since the monitoring device had to be held against the rear end of the sheep and because, as the farmer noted, 'sheep do jump about a bit', it was hard to get consistent repeated readings. The farmer could see that what ended up as a fact about contamination, started off as a messy and uncertain operation. For the farmer, the mystique and authority of other official data records began to evaporate too. According to Wynne, the credibility of expert opinion was revised, indeed re-negotiated, during the course of the farmer's experience of scientists' daily practice.

Overall, these insights from case studies in the public understanding of science reveal two main sorts of conclusions. First, they indicate that there is no one formula for transmitting scientific knowledge. The credibility of experts is in a sense always being negotiated and evaluated. A means of deploying expertise in one social context may not work in another. This is because public trust in expertise is a (perhaps *the*) central issue in PUS, and trust cannot readily be routinised. Second – and more radically – it is not accurate or appropriate to regard the public understanding of science as a one-way traffic. For example, persistent public worries about nuclear safety, coupled with the work of concerned scientists, have led to closer attention to the subtle biological pathways through which radioactive contamination can be concentrated (Wynne, 1992a: 290–5). Similarly, patients groups who have insiders' knowledge of a disease or disorder have contributed to understanding how to manage their condition, particularly on how to manage it in the light of the varied and unpredictable demands of everyday family life (Lambert and Rose, 1996). In these ways, lay publics can be active participants in the generation of new knowledge and the overthrow of old scientific beliefs. The relationship between scientific expertise and the public is far more complex than is typically recognised in calls for 'public understanding' which emanate from the scientific establishment.

A SCIENCE-STUDIES VIEW OF THE 'PUBLIC UNDERSTANDING OF SCIENCE'

In contrast to the mainstream tradition of PUS studies, I suggest that a science-studies view would lead to three key insights into the public's understandings of science and of scientific expertise. The first insight is that the public's understanding of science is not so much a question of whether people *understand* pieces of science, as a matter of the public's evaluation of the institutions of science with which they have to deal. As various case studies – including the Cumbrian sheep farmers – made clear, people not certified as experts repeatedly demonstrated an ability to learn about the intricacies of pressing technical matters with surprising speed. When Irwin et al. (1996) studied the case of residents living close to potentially hazardous chemical plants and similar factories, they found that – though many residents no doubt had an interest in the commercial well-being of the factory (because they or their relatives had jobs there or because their business thrived on trade with the factory's employees) – they also had an interest in ensuring that the plant was operated with as little prejudice to their health as possible. By contrast, factory managers characteristically had a contrasting balance of interests, judging the costs and benefits of investments in safety differently. The technical staffs who had the fullest access to company information and with a full-time concern with safety-related knowledge were precisely those employed by the company. Accordingly, the question of the public's responses to the scientific and technical issues involved could not realistically be considered in the terms customarily associated with the word 'understanding'. Members of the public formed an assessment of the scientific details of the plant's safety regime in the light not only of what they 'understood' about the technical information they were given or could acquire, but of how they evaluated the trustworthiness and the attitude of the technical staff. On the basis of studies such as this, one can argue that in the cases where science matters most to the public, PUS is less about whether people understand pieces of science than it is a question of how the public evaluate the institutions of science with which they are confronted. Trust in scientists and scientific institutions turns out to be central to the evaluation of expertise.

The second proposition is that publics commonly have their own knowledges as well, knowledges which may complement or rival expert conceptions of the matter in hand. People may have expertise by virtue of local knowledge or because of personal 'qualifications'. Though it sounds almost trivial, it is clear that medical patients have a certain kind of expertise about their own bodies and may have specially privileged knowledge of factors such as pain; in a sense they are experts alongside doctors. But this is not true only of occasions where people are expert because of perceptual privilege. It is demonstrated by the results of a recent study of public understandings of a computerised air pollution model in Sheffield, an industrial city in the north of England (see Yearley, 1999a). This model was of interest to analysts of PUS because the simulation was intended to provide advice about air-quality conditions to the public as well as being useful in scenario and planning work within the local authority. It turned out that, for a variety of reasons, local people paid little heed to the model and its predictions.

One of the reasons for this neglect was that people felt that their own expertise led them to conclusions at odds with the model's projections. For example, cyclists and traffic campaigners argued that on-street pollution was far more variable than the model implied; they insisted that their experience indicated that it was concentrated around buses with certain (ageing) engines and accordingly tended to be worst around those buses' routes and particularly their standing areas (Yearley, 2000: 115–16). This kind of perceptual expertise was ironically endorsed by the council employee with responsibility for the work-station mounted model. He knew the model's predictions well and could map them on to the city as he walked around in his private life; he acknowledged that the diminution of pollution away from the arterial routes seemed to him (and his nose) far sharper than the model suggested. This second factor implies both that people will use their own knowledge to assess the credibility of official experts' claims and that, under certain circumstances, it may prove beneficial to incorporate 'citizen expertise' into official knowledge, for example through organising panels of citizens as part of the peer review process for assessing the adequacy of the model.

The third insight is a little less straightforward. At its simplest, it depends on the suggestion that scientific knowledge claims with relevance for the public typically depend on auxiliary assumptions about the social world. For example, one aspect of the Sheffield air pollution model concerns point-source emissions, that is emissions from factories, power stations and the like. But the accuracy of data about such emissions depends not just on knowledge about the molecules emitted but also on behavioural (one might say social science) assumptions about plant managers' and operatives' conduct. It matters how well the plant is maintained, whether emissions regulations are adhered to, and so on. Yet this behaviour (unlike the behaviour of polluting molecules) is not examined in great empirical detail at all in the model, even though it is central to the overall accuracy of the outputs treated within the model. Furthermore, local residents claimed to have knowledge about the practicalities of plant management by virtue of 'inside' information from relatives and friends who worked at the plant, knowledge which directly contradicted the benign assumptions about well-behaved plant operators behind the modelling practice. For instance, local residents suggested that emissions were less carefully regulated at night and at other times when inspections were known not to occur (Yearley, 1999a: 860–1).

This point has been expressed by Irwin and Wynne as follows: they say that 'science offers a framework which is unavoidably social as well as technical since in public domains scientific knowledge embodies implicit models or assumptions about the social world' (1996: 2–3). In other words, in public contexts scientific assertions often depend on unexamined assumptions about the social world – how people will use a product or how regulators will perform. The far-reaching nature of this finding about the contextual public understanding of science can be illustrated through another example from a very different case. Epstein recently conducted a well-known study of AIDS activism around drug trials (1995). Put most simply, representatives of the US gay community demanded some say in the way that the research agenda and the drug trials were set up. Of particular relevance here, community representatives argued that conventional trials were often useless because desperately ill patients invalidated them. Knowing that some people were receiving the novel drug and that others were in a control group, patients who were supposed to be in separate cohorts got together to pool their medicine and redistribute it so that everyone stood a chance of getting at least some of the potentially beneficial drug. In this case, the implicit sociological assumption behind the testing protocol was massively mistaken. The official experts were no longer fully expert, though not so much through their inattentiveness as through the deliberate non-compliance of human subjects (1995: 421–2). In this case the expression 'the public understanding of science' is very wide of the mark since the critical issue is the scientific establishment's invalid assumptions about behaviour in the community from which their subjects are drawn.

CONCLUDING DISCUSSION

It has become clear from the discussion in this chapter that the alternative view of scientific knowledge and of expertise developed in the earlier chapters and summarised in Chapter 7 allows new light to be thrown on the current sociological phenomenon of public disquiet with the claims of scientific experts (see also Wynne, 2001). A conventional view of scientific knowledge leads towards a 'deficit' model of public understandings and focuses analytic attention on the question of why the public falls short in its comprehension. This alternative view, inspired by the ideas of symmetry and impartiality, seeks to appreciate the public's contextual understanding of scientific claims. Some expert claims are not so much misunderstood as rejected. And, in some cases, public groups appear to have counter-expertise which can support their views against those of the officially sanctioned experts. Furthermore, this alternative view implies also that the most interesting aspects of the public's understanding of science cannot be measured very meaningfully by a survey. Of course, 'variables' such as trust are central to PUS but these are not fixed dispositions. Rather, they are contextual and fluid.

As has been argued, the results of science-studies-influenced work on PUS can be elaborated into the form of three 'theorems' about the public's understandings of science:

1. The public's understanding of science is not so much a question of whether people *understand* pieces of science as a matter of the public's evaluation of the institutions of science with which they have to deal.
2. Publics commonly have their own knowledges too, knowledges which may complement or rival expert conceptions of the matter in hand.
3. 'Technical' understandings of science in public typically trade on a tacit or naïve sociology since in public domains scientific knowledge embodies implicit models or assumptions about the social world.

Though each point will not be applicable in every case of science in public, if one takes the BSE issue as an example it appears that at least the first and third points came into play. The authorities' conflicting responsibilities to protect consumer safety and to maintain the profitability of the farming and related industries meant that questions of trust were inseparable from the assessment of official advice (see also Jasanoff, 1997). The public's response was not affected only by understandings of the new, rare form of illness but also by an assessment of how trustworthy scientific expertise could be when the scientists were employed, at least in part, by an agency dedicated to promoting the farming industry. Equally, when practical steps were taken to regulate the kinds of meat available for human consumption, consumers were quick to see and be notified by consumer advocates that the proposed measures depended on a naïve sociology of the slaughterhouse. Slaughterhouse operators were told to strip out the parts of the carcass thought to be most dangerous. But how practicable it was to remove these body parts intact and how painstaking managers and workers would be was far less clear. The importance of a sophisticated public understanding of prions (the infective agent) was much less than public interpretations of the trustworthiness of advisory institutions and the practicalities of abattoir management. In this case, public scepticism indicated an unease about surrendering regulatory control entirely to official agencies. Put another way, the public saw the potential for the missing masses of ill health to slip through the government's regulatory net.

None the less, the general credibility of experts relates not only to matters concerning the public's responses, but to the social roles undertaken by experts and to the interaction between experts and other contemporary institutions. Particularly in the USA, the credibility of scientific experts has come under attack in the courts in the last two decades. Using the institutions of cross-examination, lawyers have proved adept at raising doubts about the certainty and credibility of expert witnesses, as was amply demonstrated by the internationally famous O.J. Simpson case. The points raised in this discussion will feed directly into the analysis of risks in Chapter 9 and of science and law in the subsequent chapter.

---

[1]The government of the Irish Republic, though briefly attracted to nuclear energy in the 1970s, has never had a nuclear programme and in recent years has typically adopted an anti-nuclear stance. The Irish government is concerned about emissions from nuclear power stations on the west coast of Britain and about contamination of the Irish Sea. There are no nuclear power stations in Northern Ireland either.

[2]In the contentious context of Northern Ireland, names for political identities carry heavy burdens of meaning. To be 'Irish' is generally to be in favour of Irish unity while to be 'British' is to favour Northern Ireland's union with Great Britain. To pick 'Ulster' is typically to emphasise the distinctiveness of Northern Ireland (and thus to oppose Irish unity) while the label 'Northern Irish' is closer to a neutral term. However, Irish nationalists would often reject 'Northern Irish' and say north Irish since the former alludes to the name of a political entity which they reject.

9 *Figuring out Risks*

INTRODUCTION: RISK, SCIENCE AND SOCIAL THEORY

Risk is the one topic from within the core traditions of science studies that is now widely discussed in the broader sociology and social theory literature. The aim of this chapter is to show how an understanding of risk from the point of view of science studies deepens the general sociological interpretation of riskiness and indeed leads us to reconsider the views of some leading theorists. It is clear from the outset that the connection between the language of risk and that of science is intimate. Society's practices in relation to risk typically express the idea that risks can be measured and weighed objectively. Objective analyses of risk are offered as the appropriate way of assessing an individual's or a group's exposure to the likelihood of misfortune and as the way of working out which are the risks to be most worried about. The scientific assessment of risk should allow us to determine which people are most at risk from, say, motor vehicle accidents, and also permit the consumer to figure out which are the least risky forms of transport. It is in this claim to objectivity, as well as in the ensuing disputes over how to ensure that objectivity, that the concerns of science and risk analysis meet and it is here that social studies of science has the most to say to sociological writers on risk.

This is not to imply that these are the only interesting social scientific questions about risk. For example, by developing Durkheim's celebrated suggestion that religious cosmologies reflect social structures, Douglas came to the view that cultural interpretations of the characteristics of nature indicate as much about the reflected character of society as about the underlying features of nature itself. Carefree, individualistic, dynamic cultures tend to view nature as resilient and able to look after itself; cultures which are precarious or which worry about protecting their boundaries tend to view nature as fragile and in need of protection. Douglas subsequently developed this approach for the examination of risk in advanced industrial societies. A society's risk anxieties, on her view, relate as much to the cultural 'insecurities' of that society as to the actual extent of hazards (Douglas and Wildavsky, 1982). Equally, there have been perceptive studies of the emergence of the concepts of risk, chance and probability and of the practices (such as gambling and insurance) with which they have been accompanied (see for example Hacking, 1990 and Porter, 1986). And probably most well known, there are the claims – now closely associated with Beck (1992) and Giddens (2002) – about present-day societies as 'risk-societies', meaning that society's prime concerns have shifted from the production and distribution of goods to the regulation and allocation of 'bads' (such as pollution and the threat of chemical or nuclear contamination). At one level, sociologists have discerned an historical trend towards the taming of risk, both in the sense that risk and probability have become better understood and that certain forms of exposure to risk have been reduced. Nineteenth- and early twentieth-century confidence in progress and control over nature suggested both that risks were diminishing – diseases could be controlled, dangers to food supply countered through scientific farm management, and so on – and that better understanding of risk and chance was possible. But this modernist confidence about the prospects for the control of risk has been moderated and to some extent undermined by subsequent developments. For one thing, twentieth-century technologies were associated with new forms of risk, including risks to the whole ecosystem. It seemed that the forces of progression could lead to novel hazards as well as to enhanced security.

Moreover, this process is accompanied by what may be called the 'humanisation of nature' (see Beck, 1995: 55). We have exchanged uncontrollable risks in the natural world for risky technologies whose safety crucially depends on how well they are designed, operated and run. Thus, though advanced societies may be largely freed from dependence on the vagaries of the weather, they are now dependent for their security on the good behaviour of the operators of nuclear power stations and of the institutions which guard them. It is the human regulators of risky technologies as much as the risks of nature that are to be feared. At the same time, the legal frameworks and intellectual tools developed for the regulation of risk made it possible to elaborate arguments about risk in ever more sophisticated ways, ironically engendering a more vivid appreciation of risk. This chapter will set out from the first of the issues listed above, the analysis of the 'objectivity' of risk. But arising from that will be insights into the later issues of the humanisation of nature and the characterisation of present-day societies as risk-conscious ways of living.

RISK ASSESSMENT: REGULATORS AND RISKS

As described above, one systematic and extensive literature on risk grew out of concerns to make policy objectively and to legislate for the public's exposure to risk; this literature can be categorised as concerned with 'risk assessment'. Given that risks are ubiquitous and cannot be eliminated entirely, the question for regulators and other political authorities was customarily thought of as: 'How safe is safe enough?'[1] Governments and official agencies recognise that train crashes regrettably occur, that car drivers are daily involved in collisions, that releases will sometimes arise even from well-managed chemical plants and that farm chemicals may have an impact on the environment, agricultural workers or even, occasionally, consumers. Given that no complex system can be guaranteed to be perfectly safe, the leading approach – couched in the language of economics – was to ask about the price of buying additional safety. Calculations were used to try to figure out the likely consequences of different policy choices. And, on this 'consequentialist' view, risk is thought of as the mathematical product of the likelihood of the hazard occurring and the costs (that is, the harms) to which the hazard would give rise. Looked at in this light, a large but unlikely hazard would be viewed as equally risky as a more probable though slighter harm. The policy-maker's art is then to minimise risk within the budget that society is willing to spend on safety. Some risks will inevitably remain and these are the ones which (it is assumed) society judges it is worth bearing.

In practical terms, the task of assessing risk was always much more difficult than one might suppose, and certainly more intractable than economists routinely implied. Science-studies-related work on risk assessment has indicated that, in important ways, the objectivity of the risk calculations was inescapably more precarious than its champions maintained. In part this difficulty arose because of the inherent problems of trying to harmonise risk data across various fields. It may be possible to compare the costs and risks of various systems for relating railway signals to automated train-braking systems, but transport, industrial, medical and agricultural risks cannot practically be brought within the same calculus. Records will be kept in varying ways, the commonest forms of injury or harm are likely to vary from one sector to another, and compensation costs will be worked out according to differing conventions.

Worse still, both aspects of the key calculation – the likelihood of the problem occurring and its predicted consequences – commonly defy exact specification. While there are good data on the medical and related risks of typical motor accidents on US and European highways, the probability and the consequences of nuclear power station incidents can only be calculated in hypothetical ways. There have, thankfully, been relatively few large-scale problems with nuclear installations, so frequency data are not dependable or robust in a statistical sense. Equally, the long-term consequences of various forms of incidents which might take place at nuclear installations are not yet known. In some cases, we may not have had enough time to track the long-term consequences exhaustively; in any event, the military and nuclear energy officials may well have not been entirely open about what those consequences are. In such cases, neither figure (neither the likelihood nor the consequence) is well established so that the risks cannot be calculated with any strong sense of objectivity at all. Accordingly, the comparison with the risks of other technologies cannot be objectively made either. The overall objectivity of major risk assessments appears inevitably dubious.

Nor is this problem confined to relatively uncommon installations such as nuclear power stations (of which there are only hundreds in the whole world). The risks and costs of Mad Cow Disease (BSE), an unprecedented form of infection apparently spread through the food chain to human cattle-consumers, cannot be subjected to the risk calculus in the standard way either. For one thing, little is known about the incidence of 'Mad Cow'-related disorders in human populations. Well-accepted tests for the human variant of the encephalopathy in living patients are not available. In any event, people may not wish to take any tests that could be devised if there is little prospect of successful treatment for the condition; there would be little benefit for many people in knowing that they are going to come down with an incurable, fatal brain-rotting illness. Accordingly, the spread of the disorder in the human population is unknown. As for the future consequences of the spread of the infection, it is currently unclear whether medical interventions can be developed; without such interventions the consequences could be very bad, with them the consequences might be less terrible. Again, neither the likelihood nor the consequences are known. In the absence of good data, all calculations of risks in such cases will be 'rough and ready'.

Finally, as if the preceding difficulties were not bad enough for the official approach to risk, it is not even clear that there is a single 'currency' into which all sorts of harms can be converted for the purposes of cost-benefit calculations (Stirling and Mayer, 1999: 10). We may be tempted to agree that all deaths are equally bad but even then, in practice, the deaths of the very elderly and the extremely young are not treated uniformly with those of the rest of the population. Deaths aside, there is little agreement on the 'costs' of various forms of injury and impairment. It is just not possible to come to an objective solution to the question of how many severe injuries are equal to an average death. Thus, the claims to objectivity of the standard cost-benefit approach appear to suffer from extreme limitations except in artificially limited circumstances (where the riskiness of one make of car is compared to another approximately similar marque, say, and even here the results are not beyond contestation).

Rather bafflingly in light of these persistent difficulties with formal risk-assessment techniques, one leading line of social-scientific and policy analysis in relation to risk has been devoted to exploring what are thought to be the *public's shortcomings* in regard to risk understandings. Viewed from the perspective of advocates of the objectivity of standard risk-assessment techniques, lay people and their (self-appointed) spokespersons in the media often appear to make irrational and statistically unsupported assessments of the relative risks of different hazards and to demand high levels of risk reduction without being aware of the costs. The official view has often been, for example, that ordinary people seem to worry to an unreasonable degree about the risks of nuclear power when the lives of the general public are much more at risk from motor accidents than from the operations of the nuclear industry. In the extreme, the public appears unwilling even to accept the cost-benefit approach but has no systematic alternative with which to replace it. After railway or air accidents for example, claims are often put forward about the need to enhance safety significantly. Experts typically counter that appreciable extra safety could only be bought at too high a price, a price that would make travel prohibitively expensive or that would render alternatives more risky that are (such as car travel) unreasonably cheap by comparison. As an added complication, despite frequent public protestations of anxieties about risk, market mechanisms seem to imply that people are not necessarily very risk averse: the automobile market has tended to sell on aesthetic and performance criteria rather than on safety; consumers continue to eat fatty foods and to avoid exercise despite the well-publicised risks of heart disease. The expert risk-assessment community has tended to regard its own approach as the only one with a systematic orientation towards objectivity and to play down the public's apparent dissent.

The discrepancies between public and expert versions of risk have given rise to a series of studies aimed at clarifying the basis of the public's risk perceptions (see Slovic, 1992, for example). In such studies psychometric researchers have shown a difference between people's perception of a risk which may be particularly hazardous to them (because of their lifestyle or occupation) and their generalised perception of the seriousness of risks to which everyone is exposed. Moreover, there appear to be significant differences between people's assessment of those risks to which they expose themselves and those to which they believe they are subjected by others. People seem more likely to demand that trains are safely driven than they are to constrain themselves to drive their cars cautiously.

The resulting tension between expert and public interpretations of risk has caused a problem for policy-makers. If they base regulations on expert methods and judgements, policies may be unpopular or even subverted, whereas basing policies on the public's apparent preferences threatens to make regulations arbitrary or unscientific. If the citizenry really is more tolerant of self-imposed risks than of risks visited on them by others, then there is an argument for reflecting this in public policy whatever the 'actual' exposure to risk. Similarly if, as Slovic reports (1992: 121), the public is more concerned about certain 'dread' risks (such as those from nuclear radiation) than other risks which experts hold to be of equal dangerousness, then maybe the cost-benefit equation has to be widened to take into account people's manifest preferences. Regulators find themselves confronting a tension familiar to liberal democracies: that between people's 'revealed' preferences and the recommendations supported by expert opinion.

RISK EXPERTISE: THE REFLEXIVITY OF RISK

These problems, which appear ineradicable from the risk-assessment approach, have become particularly apparent in specific institutional contexts. Risk assessments have been developed, particularly by official regulatory agencies, so as to apply to statistically representative or in some other sense 'average' people. Calculations of how long it takes to evacuate an aircraft – and thus of the requisite number of emergency exits, the width of aisles and so on – are supported by evacuation trials, but these in turn depend on notions of what constitutes a typical passenger cohort. A standard has to be established for how youthful and active, even how well shod, the body of passengers can be expected to be. Similarly, crash test dummies for assessing motor vehicle safety in the USA typically weigh around 78 kilograms, reflecting the 'standard-size' American male. The representativeness of these stand-ins became disputed in the context of the introduction of air-bags as a vehicle safety feature. Some drivers and passengers, who were shorter or lighter than the average, reported injuries arising from the inflation of the bag; it struck them with the force needed to restrain a 'standard-size' adult male and caused them harm, sometimes very severely. Now it appears, as reported in the *New Scientist* (30 March 2002: 9), that car users who are heavier than the average male also suffer by comparison with the standard motorist. From statistical data concerning crashes, it appears that heavier car users' average injuries are considerably worse than those of lighter adults, perhaps because the seatbelts restrain them less effectively in relation to the design of the vehicle interior. The measured safety of a car, intended to be an objective assessment of its safety characteristics, may not correspond to its performance for many classes of potential user. Furthermore, a lighter or a heavier car user might even find that the safety ranking of different brands of car would not correspond to the vehicles' actual performance for a person of their weight; the car measured as most safe for the average user would not necessarily be the safest for relatively overweight users.

The case is just the same with arguments about exposure to pollutants; the authorities on risk have had to construct some notion of the average person as the unit for measuring the at-risk population. Yet this ideal type hardly exists in reality and may obscure threats to specific sub-populations. Women may be different from men; pregnant women are more clearly different. The young may differ from the elderly, the house-bound from the active, and so on. These differences may prove to be far from 'academic' in particular contexts of exposure. For example, people who live near one of the UK's most controversial nuclear sites, Sellafield in north-west England, and who happen to favour a shellfish-rich diet may be exposed to a greater nuclear hazard than the average resident or average UK citizen because of the way in which shellfish filter material from sea water. The shellfish seem, in effect, to concentrate the contamination. In this case, a behavioural choice apparently unrelated to the source of the risk (from the nuclear power station and the waste treatment facility) intensifies other risk factors. In other cases, numerous background factors may act in synergy for people who live in heavily industrialised neighbourhoods or other areas of pollution concentration. Typically, the residents of such areas will be poor or socially disadvantaged, as in the case of American inner-city minority populations who may be exposed to toxic pollutants from multiple sources (see Bullard, 1994). In neither the UK nor the US case are standard risk-assessment methodologies designed to compute the risk to specially vulnerable populations.

Ironically, therefore, the great increase in formal examinations of risks and the way they are calculated has not so much diminished, as contributed to the exacerbation of risk concerns. Given the complexities of risk-assessment techniques and the high stakes involved (which might extend to the closure of a power plant or the withdrawal of important industrial chemicals), it is understandable that risk-assessment methodologies have been subject to legal and other formal challenges (see Jasanoff, 1990: 193–207). These proceedings have subjected risk assessments to critical deconstruction, questioning the precise basis for the choice of methodologies. Where official assessment methods have turned out to have consequences injurious to some but not all sections of the population, the procedure has been made to appear discriminatory. Environmental justice movements have formed to press the point that risk assessments of individual chemicals are not equivalent to a fair demonstration of the impact of cocktails of pollutants on actual citizens. Communities which experience extensive hazardous exposures – each element of which may be subject to risk assessment on its own – argue that the overall impact of multiple sources is not well gauged by standard methodologies.

Furthermore, as Jasanoff (1990: 49–57) has demonstrated, this process of critical review has been particularly extensive in the USA. Since most official risk assessments have been performed by the Environmental Protection Agency (EPA) and other branches of the Executive, their judgements have been potentially open to legal challenge and judicial review. In other words, firms and other bodies have been able to ask the courts to assess the fairness and reasonability of the executive agencies' ruling. Combined with the USA's adversarial cross-examination system (see Chapter 10), this has meant that extensive financial and intellectual resources have been directed at deconstructing risk assessments. Given all the complexities outlined above (the hypothetical nature of many harms, the inadequacy of the data on how frequently problems occur, the difficulty of identifying an average case, and so on), many risk assessments have not stood up well to judicial scrutiny. If anything, problems have been compounded by acknowledged difficulties with the experimental testing of many putatively harmful substances. As it is not feasible to test potentially damaging substances on human subjects, official agencies have had to rely on animal experiments. But different animals may respond differently to the same substances so that the suitability of the experimental animals as stand-ins for humans has also been questioned. Criticised, so to speak, both from 'right' and 'left' (from businesses which fear excessive regulation and from citizen groups who suspect that risk assessments under-estimate disadvantaged groups' exposure), the outcomes of risk-assessment procedures come to appear more and more clearly as social constructions. Commentators working from science-studies backgrounds have concluded that there is no prospect of finding an incontestable scientific basis for defending particular risk assessments (see Jasanoff, 1990: 229).

A TYPOLOGY OF RISK KNOWLEDGES

Research anchored in the science-studies tradition has examined the character of the knowledge brought to risk-measurement exercises in a different way from standard exponents of risk assessment, particularly in relation to the construction of 'certainty' and 'uncertainty' in risk calculations. For example, Wynne (1992b) has argued that, in practice, risks and probabilities are made up of many kinds of not-knowing. In some cases, it is possible to establish a hierarchy of uncertainty, as between 'risks' and 'uncertainties': with risks one knows the odds, with uncertainties only the general parameters of the problem. It is with risks of this sort, at least under the best imaginable circumstances, that the assumptions behind risk-assessment procedures are most closely realised. Most practical questions which science-in-public has to face, however, involve an additional kind of non-certainty. This Wynne terms ignorance. Ignorance refers to aspects of a problem which are bracketed off and commonly not further investigated; these are often cases where – as it is said – one doesn't know what one doesn't know.

An example will clarify the point here. When trying to work out the risks involved in climate change, the predominant risk-assessment procedure looks at the predicted rise in sea level, the more-frequent storms and floods, the anticipated impact on agricultural productivity, and so on. It then compares these harms with any likely benefits arising: warmer winters in some areas should cut heating bills and may even reduce the winter death rate. If measures to combat climate change demanded that we used much less fossil fuel and this – in turn – resulted in slower economic growth, the calculation would also try to figure out the benefits forgone as a result of the reduced economic well-being. This type of exercise is, in essence, what the Intergovernmental Panel on Climate Change has sought to do in its socio-economic assessments.[2] Of course, all such calculations contain a good deal of uncertainty: forecasts of sea-level rise are acknowledged to be inexact as are estimates of changes in agricultural output (see Zehr, 2000). But the standard calculation implicitly assumes that the biosphere continues to operate much as usual, except that it is a somewhat warmer and more energetic system. However, some scientists have proposed that one result of oceanic warming might be that the marine currents themselves would change drastically, leading to severe alterations in weather patterns. Were such a thing to happen, the customary calculations would be pitifully inaccurate. But, as yet, no one knows whether such drastic changes may occur; this would be an example of 'ignorance'. Matters covered by ignorance typically lie outside the disciplinary paradigm within which standard assessments of risk are conducted and are thus in a sense necessarily, rather than perniciously, excluded from day-to-day calculations and assessments. Nonetheless, ignorance in this sense is a different form of not-knowing from mere uncertainty; it is not adequately captured by being treated as simply extreme uncertainty.

In principle at least, more knowledge might assist in handling these kinds of not-knowing. Uncertainties might be turned into risks. New understandings might clarify specific areas of former ignorance, though – of course – there is no prospect of ignorance being overcome in general (see Yearley, 2000: 111–12). But, in addition, Wynne argues that there is a fourth consideration, 'indeterminacy', resulting from 'real open-endedness in the sense that outcomes depend on how intermediate actors will behave' (Wynne, 1992b: 117). In other words, the validity of risk assessments of systems with an organisational or human component are deeply dependent on how the systems are operated. Accordingly, conventional risk-assessment practices typically depend on unexamined and untested sociological hypotheses about those social practices that are central to the risk-producing activities. Thus, as explored in the last chapter, evaluations of the risk to local people from potentially hazardous industrial plant do not depend only on the toxicity of possible emissions (difficult to ascertain though that may be), nor even on aspects of scientific ignorance about the possible consequences of releases, but every bit as much on the behaviour of plant managers and operators. Equally, as described at the close of Chapter 8, the risks to meat consumers posed by Mad Cow Disease are governed by uncertainty and ignorance about the mechanism of infection and spread of the disease but equally by indeterminacies in the practicalities of slaughterhouse operation – whether spinal cords and other highly infective materials are actually removed according to the specified regulatory directives.

Two points about the public reception of risk expertise follow from these claims (Wynne, 1989; 1992b). First, Wynne observes that government-appointed experts, when faced with the need or chance to regulate in an area of non-certainty, are tempted to handle all forms of not-knowing as statistically treatable uncertainty, even though things of which they are ignorant cannot – by definition – be quantified. He suggests that members of the public may be sensitive to ignorance; he argues that campaigners and concerned lay people use the way that experts respond to ignorance as a yardstick when deciding to whom to extend trust and confidence. Second, he argues that the public may often be significantly more insightful than the supposed experts in relation to the matters covered by indeterminacy. His view is that, in so far as expert assessments depend on assumptions about particular social, cultural, or occupational practices of lay groups, it is likely that these publics will be more expert in these matters than technical 'experts' more distant from the relevant experiential insights.

In sum, though an ideal of objective risk has been promoted by scientists and by scientific attitudes to the valuation of the natural world, it is clear that standard risk-assessment practices are mired in troubles. Most of the figures which are fed into risk assessments are nowhere near as 'objective' as enthusiasts for the procedure assume and areas of ignorance are often passed over. On top of these considerations, there is no binding scientific reason why members of the public should accept the fundamental underlying equation, that risk equals harm multiplied by probability. In response to such difficulties, official agencies are commonly left with no alternative but to demand 'more and better' science; yet there are few grounds for thinking that further steps down the same path will resolve the problems outlined above (see Jasanoff, 1999).

CONCLUSION: RISK CULTURES

Beck and other authors have taken this impasse as symptomatic of the problems confronting contemporary society as it comes to terms with 'reflexive modernisation' (Beck, 1992). The institutions of modernity, notably the traditions of scientific analysis and legal fact-finding, have – these authors claim – been turned upon themselves with destructive consequences. For Beck this reflexive modernisation is but one facet of the 'risk society' thesis, which portrays turn-of-the-millennium industrialised societies as unusually and overwhelmingly concerned with the regulation and distribution of risks and other 'bads'. On this view, social scientific interest in risk consists not so much in the study of societal responses to particular hazards; rather it is the key to characterising present-day society as a whole.

Beck's thesis in its widest form asserts that in 'industrial risk society' risks are man-made and cause 'self-jeopardy' (1995: 78); Giddens makes the same point when he speaks of 'manufactured risk ... created by the very impact of our developing knowledge upon the world' (2002: 26). Risks in the early modern period were external to the self-conscious control of social actors. Diseases would spread, bad weather would damage harvests, fires would consume urban areas as though under the influence of external, natural forces. Even if some of these risks were exacerbated by human interventions, the contemporary perception was that they were uncontrollable. In high-modernity, by contrast, risks such as the threat of disastrous nuclear reactor incidents are plainly the consequences of human activities. On this view, Victorian and early twentieth-century confidence about the progressive diminution of risk marks not the end of risk but the transition from apparently external to societally-induced risk.

Where Giddens draws a simple division between modern manufactured risks and earlier external risks, Beck makes a three-fold distinction locating the risks of 'classical industrial society' in the middle. In this in-between category he includes the risks arising from occupational dangers in the workplace and traffic accidents. For him, the risks characteristic of a risk society are far-reaching (typically global) and beyond remedy in terms of insurance or compensation:

> At least a threefold disjunction separates large-scale ecological, nuclear, chemical and genetic hazards from the (enduring) risks of primary industrialization: first, the former cannot be delimited whether spatially, temporally or socially, and thus affect not only producers and consumers but also (in the limiting case) all other 'third parties', including those as yet unborn; second, they cannot be attributed in accordance with the rules of causality, guilt, liability; third, in so far as they cannot be compensated (because they are irreversible and global) according to the current rule of 'polluter pays', they are irremediable hazards imposed upon the alarmed safety consciousness of citizens. The calculus of risk, upon which the administration of hazards founds its rationality and safety guarantees, accordingly fails. (1995: 76–7)

Beck accompanies these claims with far-reaching assertions about the changing role and status of science. An analysis of science and technology is clearly central to the understanding of the development of industrial risk society because modern risks are typically the result of technological ventures (for example, nuclear power or the ozone-depleting chemicals in the earth's atmosphere); science and technology are involved in the cause, the diagnosis, and, with luck, the eventual rectification of the problem. He diagnoses a crisis for science and technology which accompanies the novel risk-consciousness of risk societies.

Beck is adamant about the problems confronting science: 'a momentous demonopolization of scientific knowledge claims comes about' (1992: 156); yet he is much less clear about the exact details of this crisis. His account appears to have three components. First, he advances the idea that the investigative potential of science is turned upon itself so that the 'expansion of science presupposes and conducts a critique of science and the existing practice of experts in a period when science concentrates on science' (1992: 156). Systematic analysis of science draws attention to its weaknesses, as is perhaps indicated by the kinds of philosophical analyses reviewed in Chapter 1. Second, he suggests that science is simply unable to offer much effective reassurance or assistance if risks are global and irremediable. No matter how much we know and understand, science is little help during a global catastrophe. Finally, scientific knowledge comes to be revealed as inadequate for the successful and authoritative analysis of risk. Beck makes this point somewhat opaquely:

> as science becomes more differentiated, the flood of conditional, uncertain and detached detailed results increases and becomes impossible to survey. This hyper-complexity of hypothetical knowledge can no longer be mastered by mechanical testing rules. Even substitute criteria such as reputation, type and place of publication, institutional basis also fail. (1992: 157)

Scientific claims come to be more and more open to challenge and surrogate indicators of scientific quality (such as, at which university a scientist is based) become less and less useful.

Beck's overall analytical claim has been greeted with widespread enthusiasm by many social scientists and the term 'risk society' has been popularly adopted. However, I consider that his catalogue of the difficulties and shortcomings of the language of risk is less accurate and less detailed than the account generated by science studies. For one thing, his analysis of the 'crisis' facing science is over-general and incorrect. He is right that standard risk-assessment techniques would not apply well to nuclear disasters or to global warming as we have already seen, but he under-estimates the crisis facing 'objective', scientific approaches to risk even when they are applied to the risks of 'classical industrial society'. Standard risk assessments run into problems with car seatbelts, with automated train-braking systems, with exposures to multiple industrial chemicals. He fails to see that problems of ignorance and indeterminacy beset risk assessments across the board. Second, by focusing on science's investigation of science itself, he misses the critical role of the courts and other modern institutions in exposing the limits of scientific risk assessments. Reflexive modernity in science comes about less through science-analysts reflecting on science than through big business and campaign groups challenging scientific judgements in the courts and through the media.

If his work is misleading about the crisis of science, it is also wide of the mark with regard to the characterisation of risks. For one thing, it is unclear how 'late-modern' all present-day risks are. The outbreak of Mad Cow Disease – eventually transmitted to humans – is thought to have arisen from the low-technology business of producing cattle feed from animal protein, specifically in the context of energy-saving, low-temperature process innovations. More significantly, Beck's favoured examples – such as the risk of fall-out from the Chernobyl reactor explosion – have a primitively 'democratic' quality. On the face of it, the fall-out may descend upon the poor and wealthy alike. In that sense, the 'risk society' is everyone's problem. But as the environmental justice movement, particularly in the USA, has made clear, environmental 'bads' (even the 'large-scale ecological, nuclear [and] chemical' ones emphasised in the quote above) are still often distributed quite unequally along ethnic, gender and class lines. In particular, nuclear and chemical risks are visited much more emphatically on disadvantaged communities than on privileged ones. The hazards of the risk society are not shared as evenly as Beck implies.

This chapter has shown how a science-studies informed approach to risk allows the problems confronting standard ('objective') risk assessment to be systematically understood. This approach also allows us to refine and specify in greater detail the kinds of arguments advanced by Giddens and Beck. None the less, it should be acknowledged that Beck's overall account does helpfully remind us that the key social roles of the scientific community change over time. Beck is surely right – as already argued in Chapter 7 – that scientific expertise has a greater regulatory role in late-modern societies whereas formerly its principal role was in promoting economic and military productivity. His and Giddens' point about the humanisation of nature is key too in reminding us that many current risk anxieties relate to human control over hazardous processes and substances. Widespread concern in the early twenty-first century over the possible deliberate spread of the smallpox virus (most likely by terrorists) underlines the point that the relevant 'risk' is not so much the probability of becoming infected by the disease but the degree of confidence citizens can have in the military authorities who are guarding the few remaining samples of the virus in remote research facilities. This point about the humanisation of nature is essentially equivalent to Wynne's observation about indeterminacies: the soundness of calculations about safety depends only in part on the accuracy of the claims about the behaviour of the natural world, since such calculations are critically dependent also on the reliability of underlying assumptions about the conduct of individuals and institutions. These themes of accuracy and the reflexive testing of expert representations of the natural world will be examined further in the next chapter, which focuses on the courts' treatment of science.

---

[1]This phrasing became sufficiently well known as to appear in titles and headings; see Fischoff et al., 1978 for an early example. See also the discussion in Warner, 1992, and Jasanoff, 1986. I would like to acknowledge my thanks to Sheila Jasanoff for lengthy and stimulating exchanges concerning this topic and others that appear in this chapter.

[2]For more on the work of the IPCC, see Chapter 11.

10 *Science in Law*

INTRODUCTION

Law and science are both empirical enterprises. At base they are concerned with establishing how matters stand, and within both institutions elaborate and detailed techniques have been devised for assessing the quality of evidence. On the face of it, scientific expertise would seem likely to be of particular benefit to the legal process. Experts would be able to throw special light on matters opaque to everyday actors, including legal professionals, about – say – matches between blood groups or the chemical identification of traces of drugs. From time to time, experts ought also to be able to introduce new kinds of information to the court, for example when 'DNA fingerprinting' began to be deployed in courts in the mid-1980s (Jasanoff, 1995: 55). Over the last century procedures have been codified which govern the way that scientific and other forms of expert knowledge can be utilised in court. Ordinarily, witnesses' testimony is brought before courts because witnesses have personally seen or heard or otherwise sensed some phenomenon or event relevant to the case. Witnesses bring everyday skills which support their testimony precisely because 'anyone' in their situation would have registered the same thing. Expert witnesses typically do something different: they offer testimony which is not available to 'just anyone' but which is tied to their specific field of knowledge. Frequently, such expert testimony is received deferentially and is not challenged in the way that everyday witnesses' evidence may be. However, expert witnesses are not always treated so favourably and, indeed, the issue of what is to count as appropriate expertise in any particular case may itself come to be called into question. The aim of this chapter is to consider, using several examples, the plight of expertise in court and to argue that an understanding of scientific expertise based on science studies is the best basis for interpreting this important sociological phenomenon.

EXPERTS UNDER EXAMINATION

As was pointed out in Chapter 9, the courts have been repeatedly used as a forum for challenging scientific judgements, in that case about the estimation of risk. Given that scientific judgements are aimed at achieving objectivity, the ability of courts to question science might be interpreted as contrary to common-sense. If something is objectively correct, then one would not expect it be open to successful challenge in court, unless the legal processes were somehow erroneous or misguided. Of course, it might be that only bad science gets exposed in court; in US parlance such 'scientific' work is often referred to as 'junk science' and it would be no surprise if junk expertise crumbles in the courtroom (Foster and Huber, 1997: 17). However, this conclusion is far too restrictive. Work in science studies suggests that, in specific ways, courtroom examination turns out to be peculiarly suited for the deconstruction of scientific expertise, even of expertise regarded as reputable within the scientific community.

The UK, US and related legal systems (chiefly in the British Commonwealth) set legal hearings up in such a way as to be contests. Opposing sides aim to present their own interpretation as correct and their opponents' as flawed. Cases are adversarial. This has two immediate consequences for the role of scientific experts who are introduced as witnesses. First, they are customarily witnesses for one side or the other and – however objective they may aspire to be in their testimony – are not treated neutrally in relation to the outcome of the hearing. Second, one side has an interest in discrediting the other side's experts irrespective of the quality of the science represented. In many instances the prosecution case has to be proven 'beyond reasonable doubt', meaning that the defence only has to demonstrate the presence of reasonable doubt to avoid losing the case.

A series of case studies has elucidated the consequences of this institutional arrangement. For one thing, the person conducting the cross-examination of the scientific witness has only to throw doubt on the expert's testimony; there is no obligation on the legal examiner to propose a superior interpretation of events. Though this has some resemblance to Popperian ideals about the role of simple falsification in science, it is clear both from sociological accounts of scientific practice and from post-Popperian philosophers that, within the scientific community, having doubts about an interpretation is not sufficient routinely to damn it.[1] The scientific community is tolerant of anomalies; controversies – when they arise – are typically contests between competing theoretical interpretations each of which has strengths and weaknesses vis-à-vis the other. By definition, in controversies doubts attach to both sides. Doubts are normal. Moreover, this fault-finding propensity of interrogators can be taken to extreme lengths. Legal professionals may even use the most simplistic Popperian point about past experiments pointing only inductively (and therefore not with certainty) to what will happen in the future to try to wring an admission from the expert witness that she or he cannot be *certain* about the way an empirical test would work out on the next occasion. There is a systematic mis-match between the conventions which hold in the courtroom and those which predominate in the scientific academy. Canonical versions of scientific certainty are deployed in cross-examination to imply that any uncertainty attaching to the piece of scientific testimony in hand indicates that the testimony itself is suspect.

A second important difference from the operation of the scientific community is in relation to distrust. It is clear that scientific professionals do not trust all the claims that others make. As Collins' example of the two 'co-operating' gravity-wave-physics communities (discussed in Chapter 2) made clear, scientists may sometimes treat their peers with extreme distrust. In that case, a fear that collaborators would publicise what were seen as premature conclusions led US physicists to behave in an explicitly non-trusting manner. But this potential for distrust is inevitably limited. The results from the other group were apparently accepted at face value by the US scientists; it was only the other group's judgement about the theoretical implications of the results which was treated with extreme circumspection. The scientific community, except on specific occasions, depends on trust (Shapin, 1994). Even the 'gatekeepers' of scientific standards, journal editors and referees, assume that submitted papers are based on work actually done unless they have specific reasons to doubt the authors. Of course, distrust can break out at any point and nothing is guaranteed to be immune from distrust. None the less, the prevailing ethos is one of mutual credibility disrupted only by focused suspensions of trust. In adversarial courtroom exchanges the situation is different. One side's legal representative is motivated to act as though she or he distrusted everything about the opposing expert's case. The scientific expert is likely to be faced with challenges which do not arise in the scientific community because many things are simply unquestioned. The choice of particular chemical reagents, the commercial suppliers of specific biological specimens, the use of certain statistical techniques to treat data, these are all accepted in the community. Legal challenges which require the expert to prove that these are the best reagents, specimens or techniques are quite unlike the typical challenges within the day-to-day operation of the scientific community. The courtroom advantages of asking such non-standard questions have even been highlighted in material published for legal professionals (see Oteri et al., 1982).

Such courtroom challenges commonly lead to queries of a surprisingly mundane sort, since scientific procedures are founded on a huge raft of ordinary practices, as laboratory ethnographers and ethnomethodological analysts of science have frequently pointed out (see Knorr-Cetina, 1981: 114–21; Lynch, 1985: 35–51). Scientific practice requires that laboratory receptacles are kept uncontaminated, that scales and other measuring equipment are cleaned between uses, that machines are routinely tested to ensure their reliable operation and so on. Such activities do not make it in to the methods sections of scientific papers even though they are practical necessities. A courtroom challenge may thus focus on these practical underpinnings of science as much as on the details of scientific theory. In forensic cases, perhaps most famously in the double murder trial of (American) football star and actor O.J. Simpson (which took place in California in 1994–5), one key issue was the ability to demonstrate that the DNA and blood tests – however scientifically justified and painstakingly undertaken (and these points were, of course, challenged) – were performed on the correct samples. This requires a complex and highly documented chain of evidence, linking practices at the crime scene through to the laboratory tests – whether these are done by police agencies or private companies (on these issues and this case see Jasanoff, 1998; Lynch and Jasanoff, 1998; Lynch, 2002). That defence lawyers are not necessarily raising scepticism to a ridiculous level by taking an interest in such matters is shown by a recent British experience in a possibly even more critical field. It was known that sheep could be infected with a brain-wasting disease known as 'scrapie'. Scrapie had been known for well over a century but the concern arose that some sheep might be infected with Mad Cow Disease (BSE) rather than simply with scrapie. Scrapie-infected meat appears harmless to humans but meat containing BSE is not. A programme of testing was introduced. These tests, which would govern whether the meat of mature sheep should be withdrawn from the food chain, came to a chaotic end in 2001 when it was discovered that what had been carefully collected and labelled as samples of sheep's brains were actually cattle brains. As the BBC news report rather blandly announced this: 'Experiments were conducted in an attempt to establish whether some scrapie-infected sheep actually had BSE. Results were expected in late 2001, but last-minute DNA tests showed the scientists had mistakenly spent three years examining cow brains instead'.[2] For this reason, administrative conventions about tracking and labelling samples commonly underpin science-for-the-law; the degree of adherence to these strict conventions has proved a fertile ground for lawyers' scepticism. The issue is always: Was the conduct in this case demonstrably beyond question? Record-keeping in forensic science has had to rise to very high levels to withstand the demands of cross-examiners, levels which are atypical of academic scientific practice (Jasanoff, 1998: 725).[3]

One can see how scientific expertise may be open to attack on its routine empirical flank, but weakness is also apparent on, so to speak, the opposite side. Cross-examination may also focus on scientists' judgements. In the UK, major planning decisions are commonly referred to 'public inquiries', formal hearings based on the adversarial courtroom system. A number of studies of such inquiries have been carried out from a science-studies perspective (see Wynne, 1982; Yearley, 1989 and 1992). In one such case, scientific witnesses were called by conservationists arguing for the protection of a bogland area which a developer wanted to drain and 'harvest' for horticultural peat. The scientific witnesses were interrogated by the developer's legal representative. In line with the comments in the last paragraph above, this lawyer focused first on the routine procedures by which a chart comparing the wildlife value of different peatlands had been compiled by the official conservation agency. He sought to argue that the worth of the bog in question had been over-valued in that chart (Yearley, 1989: 429–33). But he also picked up on scientific witnesses' claims that there were special features about this particular peat bog. One witness suggested that a rare butterfly (the Large Heath Butterfly) was known to frequent, even to breed on, this bog. The lawyer's response to this claim appeared not to have been anticipated by the scientific expert since, far from taking the butterfly factor as evidence of the value of the site, the lawyer argued that this judgement, not being systematised within the chart, amounted to special pleading (Yearley, 1989: 433–5). If different scientists did not agree about the importance of some feature, then – so it was suggested – that feature could hardly be scientific. And if was not fully scientific, then the privileges attaching to scientific experts' testimony should not extend to this matter. In the extreme, scientists' judgements, precisely because they are 'judgements' and not the outcome of formulaic practices, can be presented as tendentious.

In summary, though scientific experts are granted a special position as courtroom witnesses, their authority has been much more susceptible to attack than one might have supposed it would be. Cross-examiners have been able to deconstruct scientific authority. In particular, in so far as experts' testimony depends on the performance of routine tests, cross-examiners have found they can beneficially press at every single step in that routine. And, in so far as experts' testimony depends on expert judgement, that judgement can be portrayed as 'personal', insufficiently scientific and thus potentially misleading. As Jasanoff has recently observed, this has meant that – in certain areas of forensic science where courts are the main 'market' for the knowledge produced – courtroom standards regarding the adequacy of evidence have taken over from the notions of adequacy which tend to prevail in the scientific community; in such instances, legal criteria of adequacy even inform the conduct of everyday scientific business (1995: 50–2). These difficulties experienced by scientists are comprehensible in the light of a science-studies informed understanding of science. Though scientific practice is methodical and though scientists subject each other to critical scrutiny, the institutions of science depend also on trust and judgement (as discussed in Chapter 7).[4] By constantly calling this trust into question and by treating judgement as unsystematic, adversarial cross-examination continuously threatens to deconstruct scientific authority. This mis-match between science and the law may be further exacerbated because legal processes are concerned with securing justice in the particular case whereas scientific procedures are aimed at identifying claims that are robust in the long run. In the next section, these same issues will be examined from a different vantage point through an example (the Daubert case) in which the US legal establishment was obliged to focus on the question of how to define science in order to legislate over which kinds of scientific expertise should be recognised in court.

THE BACKGROUND TO THE DAUBERT CASE

Late in the 1980s, a case was brought against the pharmaceutical company Merrell Dow by two children and their parents alleging that the children's birth defects had been caused by the fact that, during pregnancy, the mothers had taken a prescription anti-nausea drug (Bendectin) manufactured by the company. The District Court which heard the case was faced with rival claims from scientific experts supporting the two sides. The respondent, Merrell Dow, was supported by the affidavit of a 'well-credentialed expert' (United States Supreme Court 1993, cited in Foster and Huber, 1997: 277[5]). This single expert, a physician and epidemiologist, presented evidence based on a review of the extensive scientific literature on the drug which indicated that 'maternal use of Bendectin [had] not been shown to be a risk factor for human birth defects' (Foster and Huber, 1997: 277). In short, in all the 30 published studies, dealing with 130,000 patients, it had not been shown to be a teratogen. Sadly, a small proportion of children suffered birth defects and this was true whether their mothers had taken the drug or not. Disorders were not statistically associated with the use of Bendectin in a way that would suggest that the drug caused these children's problems.

The family's case was supported by a larger array of scientists, eight in all, who had qualifications and credentials generally comparable to those of Merrell Dow's expert witness. In outline, their argument was not that the studies cited by the pharmaceutical company's witness were not reputable. Rather, they claimed that they were in possession of different kinds of scientific evidence which did point to the drug's teratogenic (malformation-promoting) capacity. Their studies were founded in a variety of disciplines and approaches. Thus, they offered in vitro and live animal studies which indicated that Bendectin was associated with developmental malformations. They offered 'pharmacological studies of the chemical structure of Bendectin that purported to show similarities between the structure of the drug and that of other substances known to cause birth defects' (cited in Foster and Huber, 1997: 278). Finally, they put forward arguments from the results of unpublished re-analyses of epidemiological studies on the drug. The court thus faced a choice between competing potential scientific claims, and the judge had to decide how to proceed: should all the scientific evidence be admitted into court and presented to jurors or should the judge decide that only some of the scientists were properly qualified to act as expert witnesses in this matter.

The company pressed the court for a summary judgement on this matter and the judge, invoking the so-called Frye ruling (see below), reasoned that, because only the respondent's scientific expert was employing the recognised and 'generally accepted' basis for scientific reasoning about the cause of birth defects in this area, the petitioners' scientific evidence could not be admitted. In effect, this meant that the families had no scientific representatives while the company retained its epidemiologist; Merrell Dow would have won the case hands down. The relevant Court of Appeals generally supported this judgement. But it was clear that, alongside the petitioners' case, a larger principle might be being determined in this hearing and the question of the admissibility of scientific evidence was referred to the Supreme Court, the final arbiter for most legal issues in the US system. The Supreme Court, in both the opinion issued in the report by the majority of judges and the accompanying minority report, took a somewhat different view on the admissibility issue. To appreciate their argument it is necessary briefly to review the Frye ruling.

LEGAL INTERPRETATIONS OF ADMISSIBLE SCIENCE IN THE USA

Frye versus the United States was a case argued in 1923 which turned on the admissibility of evidence derived from a precursor to the lie-detector machine (see Foster and Huber, 1997: 279–80). The results of this lie detector were ruled inadmissible, essentially because the device and its successful operation had not yet won general acceptance in the scientific community. In the oft-quoted words of the judgement: 'the thing from which the deduction is made must be sufficiently established to have gained general acceptance in the particular field in which it belongs' (cited by the Supreme Court, 1997: 280). This argument about gaining 'general acceptance' came to be used as a general criterion for the admissibility of scientific evidence in the succeeding decades. In the Bendectin case the lower court judges had argued that epidemiology was the generally accepted method for assessing the causal influence of drugs on health defects in unborn children; the other arguments were accordingly inadmissible since they failed to meet the criterion of general acceptance in the field in question. In particular, the re-analyses of epidemiological data, being unpublished and thus not having been subject to peer review, were interpreted as conspicuously deficient as regards 'general acceptance'.

The argument made in the Supreme Court was in effect based not on a challenge to the content of the Frye rule but on the claim that this principle had been superseded by new rules of evidence, the Federal Rules of Evidence, introduced in 1975 (1997: 280). The Supreme Court justices agreed, arguing that the subsequent rules of evidence were intended to introduce a more liberal standard for admitting expertise. Hence: 'Frye made "general acceptance" the exclusive test for admitting expert scientific testimony. That austere standard, absent from and incompatible with the Federal Rules of Evidence, should not be applied in federal trials' (1997: 281). In particular they cite Rule 402 to the effect that 'All relevant evidence is admissible' unless specifically proscribed (1997: 280), and Rule 702 which states:

> If *scientific*, technical, or other specialized *knowledge* will assist the trier of fact to understand the evidence or to determine a fact in issue, a witness *qualified as an expert by knowledge*, skill, experience, training, or education, may testify thereto in the form of an opinion or otherwise. (1997: 281, emphasis added)

The Supreme Court justices were unanimous in reasoning that these rules at least allowed the possibility that other kinds of scientific evidence might be significant in the Bendectin case; they ruled that the judgement based on the Frye criterion should be overturned.

So far matters appeared rather straightforward. But the Supreme Court had re-created for itself the problem which the Frye ruling had sought to handle. Expressed in the terms of Rule of Evidence 702 (see above quote), the problem is: How does the court know (a) what is to count as 'scientific knowledge'; and (b) which putative experts are 'qualified as an expert by knowledge'. The court did not want to exclude relevant, valid scientific information through strict adherence to a restrictive rule, as with Frye; in any event Frye was hardly unambiguous since there could always be disputes over how what exactly constituted 'general acceptance'. On the other hand, courts need a way of regulating which kinds of scientific opinion are allowed since scientific (and other) experts enjoy a privileged position in court, being able to introduce evidence based on experiments and tests performed by a wide range of people not actually represented in court: 'Unlike an ordinary witness ... an expert is permitted wide latitude to offer opinions, including those that are not based on first-hand knowledge or observation' (1997: 283). In other words, expert testimony is a license to introduce specific kinds of 'hearsay' into court proceedings; it is clearly of practical importance that such license is only extended to the 'right' people.

In recognition of the importance of the issues at stake before the Supreme Court, many parties submitted *'amicus curiae'* (friend of the court) briefs aiming to advise the court on how it should conceptualise this matter and thus redraw the standard for admissibility. Among the 21 briefs submitted (Solomon and Hackett, 1996: 137), two are of particular significance to the current argument: one by the Carnegie Commission on Science, Technology and Government and the other jointly by the American Association for the Advancement of Science (AAAS) and the (US) National Academy of Sciences (NAS).

THE VIEWS OF THE 'AMICI'

The *amici*'s views are of interest both because of their impact on the Justices' final ruling and because of what they tell us about the practical value of the philosophy of science and science studies in advancing our understanding of science in court. The brief submitted by the AAAS and NAS, representatives of the US scientific establishment, is concerned that the Supreme Court understands science sufficiently well that the Justices do not lower the threshold for recognised scientific expertise too far. Detailing at length the need for disinterestedness in scientific assessment and explaining at equal length what the purpose of peer review and other associated practices are, these *amici* propose that:

> Courts should admit scientific evidence only if it reasonably conforms to scientific standards and is derived from methods that are generally accepted as valid and reliable. Such a test for admissibility would incorporate the factors, including the results of peer review, that scientists consider in evaluating each other's work. (Brief of AAAS/NAS as Amici Curiae for Respondent, 19; also cited in Solomon and Hackett, 1996: 137)

These *amici* go on to elaborate their philosophy of science (an amalgam of empiricism, Mertonian disinterestedness and Popperianism) and to spell out their advice, suggesting that judges should look for good science, for peer-reviewed science, and should decide what science to admit before a trial starts. In other words, these *amici*'s aim is to instruct judges how scientists determine good science so that judges can use the same principles to recognise proper science, ensuring that only such science will be admitted into court. Then courts will be able to privilege the evidence offered by the expert witnesses since its scientific standing will be beyond question. Exactly how they think their advice will assist judges in practice is unclear since, among their examples of what to exclude, is 'testimony in a traffic-accident case that disputes the applications of Newton's laws of motion' (Brief of AAAS/NAS as Amici Curiae for Respondent, 21). It is hard to believe that any competent judges would find such examples terribly illuminating. The lack of a better illustrative example indicates that the advice they offer is less useful as a discriminatory device than they would like to suppose.

Turning to the Carnegie Task Force submission – a neutral brief in support of neither party – this report too attempts to advise the legal authorities on what view of science to adopt. In short, the Carnegie view is that while judges cannot, and should not, be expected to determine the substantive validity of particular scientific claims advanced by experts, they can focus on the question of whether experts 'engaged in recognized forms of scientific practice' (Brief of Carnegie Commission as *Amici Curiae* for Neither Party, 5). In other words, they stress the extent to which the 'scientificness' of science resides in scientists' adherence to forms and procedures of inquiry. To determine whether 'scientific claims have been developed within the bounds of a recognizable form of scientific inquiry', they propose three criteria (Brief of Carnegie Commission as *Amici Curiae* for Neither Party, 11):

1. Is the claim being put forth testable?
2. Has the claim been empirically tested?
3. Has the testing been carried out according to a scientific methodology?

Elaborating on the first of these, the Carnegie authors invoke Popper to indicate that scientific statements are ones 'capable of being proven false through observation or experiment'; moreover, they claim that 'the data produced through this testing must be capable of replication'. The next issue, deciding whether a claim has actually been tested empirically, is treated primarily as a matter of validation by the scientific community. In other words, the authors take peer review and publication as standard indicators of the fact that a claim has been tested, although they do allow that some claims may have been tested in ways which have escaped peer review. In such cases, courts will have to look for alternative indicators. Their final question, whether testing has been carried out in accordance with a scientific methodology, is accordingly crucial. Their point here is not that there are timeless, cross-disciplinary standards of adequacy, but that, within a contested area, it is far easier to agree on methodological standards than on particular substantive claims. The Carnegie authors even suggest that courts may be able to find 'neutral' scientists, not necessarily highly knowledgeable in the field directly in question, but able to apply knowledge of general methodological standards. As an example, they cite the case of a geologist who advised a court over the adequacy of water-sampling methods in relation to claims about contamination by toxic waste. The geologist's relevant skill lay not in knowing about toxic wastes but about water sampling.

THE SUPREME COURT'S VIEWS ON ADMISSIBILITY AND THEIR CONSEQUENCES

Having considered the *amicus* briefs, the Supreme Court came to an opinion. The judgement starts off by examining what 'scientific knowledge' might mean: this is done by referring (somewhat surprisingly in the light of all the philosophical literature to which the *amici* had referred) to a dictionary definition for 'knowledge' and by suggesting that to be scientific, knowledge 'must be derived by the scientific method' (cited in Foster and Huber, 1997: 282). The Supreme Court judgement then goes on to consider the situation of a trial judge 'faced with the proffer of expert scientific testimony'; such a trial judge has to determine whether the 'reasoning or methodology underlying the [proposed] testimony is scientifically valid' (1997: 283). They go on to assert that: 'We are confident that federal judges possess the capacity to undertake this review' (1997: 283). They note that a variety of factors will rightly influence the view taken by any particular trial judge at a particular inquiry and they explicitly decline to give a 'definitive checklist or test' (1997: 284). However, they make the following four 'general observations'.

First, they propose that a key question is whether a 'theory or technique ... can be (and has been) tested' (1997: 284). They cite empiricist philosophies of science and Popper and offer a version of a falsificationist theory of science, according to which scientific knowledge is special precisely (in words quoted from Popper) because of its 'falsifiability, or refutability, or testability'. Next, they suggest that a pertinent consideration is 'whether the theory or technique has been subjected to peer review and publication' (1997: 284). Citing a variety of authors, including science-studies authors Jasanoff and Ziman,[6] they note that this consideration is far from clear-cut. The peer review system itself is fallible; it may be conservative and thus risk screening out good but innovative proposals, and some work may just be too specialised to have found a publication outlet. All the same, peer review is a good means of examining the certified quality of scientific work since scrutiny by peers is likely to reveal methodological and other weaknesses. Accordingly, peer review and publication are said to be relevant but not 'dispositive' factors in assessing the validity of a technique or methodology. Third, the Supreme Court judgement proposes that, for a particular scientific technique, the court 'should consider the known or potential rate of error ... and the existence and maintenance of standards controlling the technique's operation' (1997: 284–5). Lastly, the Justices propose also that the elderly dog of 'general acceptance' may yet have its day. Widespread acceptance can be viewed as a positive indicator while, on the other hand, a widely publicised technique which has won little support is liable to be looked at sceptically.

Though they are careful not to present these as rules or even as precise criteria, it is clear that the justices' idea is that judges should use considerations of this general sort to allow into court only knowledge-claims which are relevant and scientific, where 'scientificness' is likely to be indicated by:

1. testability
2. successful peer-review and publication
3. declared [and presumably low] error rate
4. widespread acceptance in the scientific community.

In sum, the approached adopted by the Supreme Court proposes that it is up to judges to recognise science. Judges should then admit good and relevant science into their courts. The four Daubert 'pointers' are indicative of the kinds of consideration which judges should use in making their decisions about which science is 'good'.

One very important implication of the Supreme Court's work relates to its impact on the original case. As Solomon and Hackett note (1996: 152), the lower appeal court 'affirmed its previous decision to reject the causative evidence brought by the Dauberts' since the families' evidence was judged not to satisfy considerations such as extensive peer review and widespread acceptance.[7] The lengthy inquiry had caused no change to the standing of the evidence in this case; all the same, the company had already withdrawn Bendectin. However, the ruling had far-reaching implications for subsequent interpretations of the question of admissibility since, however much the Supreme Court insisted that its four indicators are not criteria, they have become a kind of touchstone.[8] Black et al., writing in a legal journal, for example, gave them a favourable reception and referred to them as the 'Daubert test' (1994: 721).

SCIENCE STUDIES AND THE NEW 'CRITERIA'

The enthusiasm of Black et al., stems from the idea that the Daubert test does two things. It supposedly encourages judges to assess science using the same considerations that scientists use to assess it. Second, and indispensable to the first, the test accurately encapsulates how scientists do in fact make their assessments. Sadly, there are difficulties with this second proposition.

Let us briefly consider these indicative criteria one by one. It is no accident that testability occupies poll position in the list since, as Mulkay and Gilbert have noted (1981), the idea of testability has proven attractive to the public statements of many prominent scientists. On the face of it, testability looks very promising as a criterion; indeed it is difficult to imagine any scientist or scientific theory which boasted of a lack of testability. But – as indicated in Chapter 1 – there are several grave and well-known problems with the notion of testability as a distinctive feature of the scientific method. The chief among these are revealed in the apparently straightforward logic of falsification advanced by Popper. As discussed earlier, Popper's position soon ran into difficulties for two main reasons. First, it is never easy to work out whether any particular experiment should count as a test of an idea. Perfectly legitimate and successful scientists appear to ignore lots of experimental tests which seem to falsify their theories because they assume the test was poorly done and was inconclusive. Gravity-wave scientists continue to get millions of dollars to look for gravitational radiation even though most believe that all tests devised to date have been negative; they assume the tests were not sensitive enough. Second, scientists typically respond to a negative test result by revising their theory rather than by rejecting it; indeed this way of proceeding became enshrined as part of Lakatos' methodology for scientific research programmes. According to Lakatos, all good theories are very likely to have failed tests and then been revised (and improved) so as not to fail them again.

These difficulties are typically glossed over in the commentary literature; Black et al. 'demonstrate' the significance of falsification by considering the hypothesis that apples are made of iron (1994: 755). Such an hypothesis would apparently be verified by the observation that apples fall under the influence of gravity. But, they caution, this observation should not be taken as providing proof of the hypothesis. By contrast, the observation that apples float would allow us to dispense with the theory altogether. The fact that we already know that apples aren't made of iron helps to make this example appear persuasive. If, on the other hand, there were truly any doubt about the nature of the composition of apples, we might expect an observation of floating apples to be taken as indicating a new low-density form of iron, or might suspect that there was something defective about the conduct of the experiment. We know, after all, that iron things can float (boats do so most of the time) and yet that observation doesn't falsify our theories about metal since we use another theory (about the mass of displaced fluid) to explain the capacity of hollow iron craft to exhibit buoyancy.

Having devoted a good amount of time to this criterion, it is worth summarising by saying that, outside of made-up examples where we already know the 'right' answer, falsification is not a simple discriminatory device. Clearly, testability and testedness are a good idea but whether they can work as screening criteria must be very much in doubt; testability is as much an expression of an aspiration as a descriptive account of scientific practice. These doubts reinforce those of the Supreme Court's minority opinion, to the effect that judges may have some difficulty in working out exactly what falsifiability means: in the words of Chief Justice Rehnquist, 'I defer to no one in my confidence in federal judges; but I am at a loss to know what is meant when it is said that the scientific status of a theory depends on its "falsifiability", and I suspect some of them will be too' (Supreme Court minority opinion reprinted in Foster and Huber, 1997: 289). Given a scientific theory relevant to a case and a piece of evidence which supposedly tests that theory, no judge will be able to tell whether the theory is flawed or the evidence misleading; instructing the judge to use the criterion of falsifiability will not help her or him work out whether this instance is to count as a falsification or not.

Turning more briefly to the other three criteria, the first and third (testability and error rate) have in common that they appear to offer an objective standard by which scientific practice can be assessed; by contrast the second and fourth (peer review and general acceptance) relate to judgements that the scientific community makes of itself. Accordingly, I shall turn next to criterion three. If there is an established 'error rate' for a particular scientific technique then it is clearly a good idea that the court should be informed of it. But the notion that there are known error rates repeats some of the misconceptions which underlay the difficulties with testability. Science is a practical activity and practitioners have views about the relative skills and abilities of their colleagues. Thus claims about error in scientific evidence will tend to turn, at least in the most significant cases, on disputed 'errors' rather than on standardised ones. For example, 'error rates' in DNA testing might be thought to turn on the frequency of certain genetic patterns in the population (how often might some innocent person share DNA characteristics with the guilty person) and on the technical dependability of the apparatus. Instead, a good deal of concern has focused on the reliability of the workers in the various private laboratories doing the DNA testing and on the scientific legitimacy of the standardising procedures used to 'correct' for the inevitable case-by-case variations in DNA tests. The influence of these latter kinds of complicating factor cannot be measured by an error rate. In such cases one cannot arrive at an objective error rate without making assumptions about the competence and so on of the practitioners, exactly the point which may be contested in court. The apparently 'objective' response to this problem (having people declare the error rates of their procedures) threatens to sweep the problem under the carpet rather than to confront it head on.

Analysts from the social studies of science will have rather fewer difficulties with the remaining two criteria because both allow scope for the exercise of scientific judgement. Peer review is, of course, imperfect in many of the ways identified in the Supreme Court's judgement but, provided no one harbours illusions about this, there is no great difficulty with including it as one basis for judging the appropriateness of scientific evidence. Similarly, with the idea of general acceptance. This guideline appears to acknowledge a role for judgement in just the way that science studies proposes. Ironically, therefore, of the four criteria, those that are more apparently cut and dried (demanding the acknowledgement of an error rate or the use of a falsifiability check) are the least realistic, while those which specifically call for the use of judgement are more practically viable.

This realisation hints at a broader conclusion which has been helpfully expressed by Jasanoff:

> Conventional legal scholarship, with its deep-rooted commitment to the existence of objective facts, offers relatively few resources for understanding what makes, or unmakes, the credibility of scientific evidence in the courtroom. ... Evidence ceases to be acceptable in the eyes of the law when it is contaminated by preventable technical or moral failings – for example, a break in the chain of custody, unethical behaviour by a lawyer, dishonesty on the part of an expert witness or reliance on flawed science. The possibility of more radical contingency in the production of evidence lies outside the normal scope of legal analysis and self-awareness. (1998: 715–16)[9]

But social studies of science draws attention to these contingencies, contingencies which make the 'Daubert test' unrealistic and far less 'scientific' than its proponents assume. The Supreme Court didn't intend that their four indicators should be treated as rules; rather, they wished courts to be like scientists and to assess evidence in a scientific way (see Jasanoff, 2001). But though this recognises that science depends on judgement, it does not fully acknowledge the extent to which it depends on *skilled* judgement. Without the skills, courts will come unstuck. Using rule-like criteria, they may well be even worse off. Of course, the final irony is that the Daubert criteria are now released upon the world, a 'social fact' in Durkheim's sense (see Edmond, 2002). Lawyers and courts will be obliged to work with this new terminology and will grow expert in manipulating the terms' meaning. There is a 'Daubert industry'. But this seems like an odd version of progress, since the vagueness of Frye has been replaced with the misleading concreteness of Daubert.

CONCLUDING REMARKS

This chapter has demonstrated the value of science-studies work for understanding the law's interpretation of science. Two remarks will bring this chapter to a close. First, social theorists (most notably Beck) have written on reflexive modernisation, the destructive application of modernity to itself. One might be tempted to interpret the Daubert case in that light. The leading legal-rational institutions (as Weber termed them) have investigated themselves in an attempt to define the authority of science authoritatively and the project has misfired. But I suggest science studies gives a fuller account of science's problems with the law. It is not so much a problem of reflexivity as a problem of conflicting institutional designs. The adversarial system favours relentless distrust, while science – which depends on routine trust and skilled judgement – is ill equipped to withstand its scrutiny. The attempt to rescue science's authority by locating a philosophical key to its 'exceptional' character (such as falsificationism) can only fail since such rational principles also overlook trust and judgement.

Second, the issues identified in this chapter are not limited to arcane aspects of national law. Exactly analogous issues can soon be expected to become pressing for international institutions such as the World Trade Organisation (WTO). As globalisation proceeds, the WTO is committed to eliminating barriers to trade and to combating unjustified protectionism. However, in many cases the justifiability of barriers to trade turns on technical or scientific considerations. Thus, at present, genetically modified foodcrops are widely planted in North America. European governments have largely resisted, arguing that there may be environmental or (just possibly) consumer safety issues which mean the crops are unwelcome. The USA rejects these arguments, claiming that there is no scientific evidence for environmental or consumer dangers and that the Europeans are simply being protectionist. The WTO is the body required to resolve such disputes, which it does through a form of legal hearing. In such matters it is inclined to look to scientific experts for the answer to these fraught political problems, since the WTO maintains that technical barriers to trade can only be lawfully maintained if they are technically valid. But judgements about technical validity demand that the WTO is able to identify disinterested experts to rule on the matter. Over GM food, the WTO will find itself in exactly the position of the courts reviewed in this chapter; it will not find it easy to identify who the relevant experts are without, at the same time, implicitly determining the outcome of the judgement in advance. To choose which scientific expertise to admit is already to decide the argument. In the near future we can expect more problems of incompatibility between science and law, not fewer.

---

[1]Also, as pointed out in Chapter 1, Popper was often a less naïve falsificationist than the most formulaic accounts of his views might imply.

[2]Report from 10 January 2002 accessed on the BBC News website http://news.bbc.co.uk/1/hi/uk/1569739.stm on 26 January 2003.

[3]This is true not only of science in court but also science used for regulatory purposes (food safety testing for example) where protocols are specified in such minute detail that few university science departments desire or win accreditation (see Irwin et al., 1997: 26).

[4]This is very nicely illustrated in a case from Australia where a scientist went to law to try to get the court to impose limits on public presentations and fundraising by a Christian fundamentalist who had claimed to have found remains of Noah's Ark in eastern Turkey. The scientist argued that the fundamentalist was in demonstrable error and should therefore be prevented from further misleading the public under 'trading practices' legislation. The cross-examination of the scientist reveals how difficult it is to codify what is scientific about what scientists do; see Edmond and Mercer, 1999: 330–2.

[5]As readers outside the USA may find it easier to locate this source (and those inside the USA no harder), I have given page references to the reprint of the 'opinion' in Appendix B of Foster and Huber's book (1997). I would also like to acknowledge my thanks to Sheila Jasanoff for many extremely helpful conversations about this case and about the points of principle it raises.

[6]John Ziman is well known for his work on social mechanisms for ensuring the quality of scientific knowledge; see his *Reliable Knowledge* (1978).

[7]For further details of the reasons for the court's rejection of the petitioners' arguments see Foster and Huber, 1997: 255–7.

[8]For a very informative analysis of the way this 'touchstone' has been interpreted, chiefly in legal circles, see Edmond and Mercer, 2000.

[9]On the background to the idea of legal 'facts' see also Poovey, 1998.

11 *Speaking Truth to Power: Science and Policy*

INTRODUCTION: THE PROBLEM OF SCIENCE FOR POLICY

The discussion in Chapter 8 was concerned with making sense of public responses to the pronouncements of scientific experts. I argued in that chapter that a view of scientific knowledge informed by science studies gives a better account of the nature of contemporary public disquiet with expertise than conventional views. Furthermore, the approach set out there supplied an understanding of why 'publics' may have forms of expertise which can complement or challenge the expertises offered by official scientific authorities. But that discussion, by focusing on publics' understandings, left open the issue of the best way in which to analyse the role of scientific advisers to political authorities. On the face of it, as science is the best knowledge we have of how the world works, it is wholly understandable that scientists will be important providers of advice for governments and policy-makers. But the relationship between political authorities' need for advice and the generation of scientific insights is complex and indirect. From as early as the seventeenth century, members of the scientific community have insistently promoted the view that, though scientific knowledge is useful in advice-giving, knowledge which turns out to be the most useful is generally not deliberately developed in order to provide advice to officials. Studies in atmospheric chemistry began long before current concerns over air pollution and climate change, while calculations of orbits were made centuries in advance of our ability to put satellites into space. Thus, even though scientific research provides a platform for advice-giving, basic scientific research is not normally carried out in order to give advice.

Still, even if research is not typically conducted with policy ends in mind, it has long been asserted by scientists that the scientific community has two advantages in relation to the provision of advice. First, scientists often know esoteric things and are the sector of society with the most systematic and authoritative claims to knowledge of how bits of the natural world function. Second, members of the 'pure' scientific community – ideally at least – are single-mindedly committed to the advancement of scientific knowledge in their specialist area, to the pursuit of accurate and objective understanding. Their location within the basic scientific community gives them an impartiality regarding the implications of the results of their work; in relation to the two examples given above – for example – they are not insurers worried about climatic changes nor are they the owners of telecommunications satellites. They can offer sound advice precisely because they are not party to these commercial or political concerns. They are independent because they are devoted to science alone. As was discussed in the Introduction, this idea is precisely what makes the term 'pure science' so evocative.

The established literature on science policy rehearses this argument about disinterestedness and impartiality over and over. Scientific research is to be funded partly because research may lead to economic benefits, partly because it contributes to the advancement of civilisation and partly because of the policy-relevance of the knowledge produced.[1] But, despite these positive considerations, a common experience is that scientific expertise does not in fact lead to the adoption of agreed, best policies. Thus, as reviewed in the last three chapters, there has been considerable unease over expert advice about how to respond to Foot and Mouth Disease; official estimations of risk in policy areas such as nuclear safety are time and again at odds with the public's approach to these matters and have been met with widespread scepticism; and opposing sides in court commonly manage to find their own scientific experts, thereby undermining the suggestion that the scientific community is likely to generate agreed recommendations in disputed policy areas. With the scientific community's apparently impeccable credentials for advice-giving, how is it that the advice proffered is often so weak in practice?

Conventional explanations for this phenomenon tend to be of two sorts. First, there are explanations which focus on ways in which scientists' independence is compromised. Thus it may be that scientific advice is not in practice set up as disinterestedly as it would ideally be. Scientists may act as guns for hire, willing to present the kind of evidence that partisan lawyers or other advocates would wish to hear. In the same way, governments may appoint people to advisory committees who are selected precisely because they are thought likely to give the kinds of advice politicians dearly would like to receive. In some cases, the scientific community itself may even generate incentives that threaten the ideal of impartiality. It has been suggested for example that the Intergovernmental Panel on Climate Change, the international expert body which was developed to advise on the science of global warming, has an institutional interest in playing up the threat of humanly-caused atmospheric warming (this suggestion is described in Boehmer-Christiansen, 1994b: 198; see also McCright and Dunlap, 2000). In such cases, so the accusation goes, the scientific community has a material interest in the plausibility of the scientific warnings about climate change because scientists benefit from the continuation of lavish research funding on this topic. The second explanation dwells less on threats to impartiality but concentrates more on the nature of the problems to be addressed in policy-related science. Authors, following Weinberg (1972), have argued that often the policy questions to which answers are sought are not the ones that science itself asks. This was the point of the analogy (described in the Introduction) with the story of the goose that laid the golden eggs. Only when left to itself would scientific research produce results of benefit. The difficulty with policy-relevant work is that the question and the timing of the query are selected by the nature and condition of society's problems, not the state of scientific knowledge and its internal trajectory. The term 'trans-science' was accordingly coined to describe this kind of research since scientists were being asked to answer apparently science-like questions but without the circumstances being suitable for authoritatively correct solutions to be devised. As Weinberg put it: 'I propose the term trans-scientific for these questions since, though they are, epistemologically speaking, questions of fact and can be stated in the language of science, they are unanswerable by science; they transcend science' (1972: 209).

Armed with arguments of this sort, analysts in political science and various branches of policy analysis have been able to adopt an 'if-only' analysis of science advising. If only apparently reputable scientists would not act as experts for hire and if only the scientific community could overlook its self-interest as a profession, then scientists could get back to advising disinterestedly in those areas where scientific expertise was properly relevant. However, a much less optimistic view has been proposed by two analysts of science policy whose understanding of scientific advice-giving has been greatly influenced by the science studies literature; the model to which their evidence points is known as the over-critical model of science advising.

THE OVER-CRITICAL MODEL

Collingridge and Reeve (1986) argue that the 'ideal' of basic scientists as policy advisers is both inaccurate as an account of the ways things stand and also misleading as an ambition to which the scientific community and policy-makers may aspire. Essentially, they set out to demonstrate that the principal assumptions underlying the received ideal are all incorrect and unwarranted. The arguments presented in their work imply that received ideas depend on four leading assumptions. First, it is assumed that scientific researchers have autonomously elected to develop knowledge which then happens to be relevant to some question of policy. However, in many cases it turns out that quite the reverse is true. Speaking of the case of research into environmental lead contamination (from sources such as paints and from leaded vehicle fuel), they insist that this 'topic has only been researched because of its relevance to policy, for it is clear that science under autonomous control would never have investigated this little corner at this time' (1986: 35). This point is fundamentally the same as that made by Weinberg, since the reason that Collingridge and Reeve give for scientists having autonomously avoided such questions is that the consequences of lead exposure are very hard to measure. For ethical reasons, human subjects cannot be deliberately exposed to lead in experiments, and surrogate measures are full of technical difficulties. Moreover, the long-term consequences of low levels of exposure are hard to detect unambiguously. In the face of such intractable methodological difficulties scientists may have noted the question but will not have been motivated to study it in detail until they were led to it by the demands of policy-makers.

Collingridge and Reeve propose secondly that policy problems will typically not fall squarely within the confines of a single discipline; more likely, they will occur at the intersection of different disciplines' concerns. Rather than assuming that perspectives from different disciplines are likely to align, Collingridge and Reeve argue that 'Science is not one but many' (1986: 22). Researchers from different disciplines will be accustomed to communicating primarily within their own discipline. They will have established ways of conducting their work which may not be in conformity with the practices common in other disciplines. They may even take different analytic assumptions for granted. In the lead-exposure case, geochemists and industrial-health workers arrived at different measures of background lead levels against which exposures should be compared. Collingridge and Reeve also use the case of UK policy debates over educational policy and the heritability of intellectual ability as a second example (1986: 89–95). The disciplines of psychology and genetics approached the calculation of the heritability of intelligence in contrasting ways, which meant that there was no single expert voice to which educational policy-makers could turn.

The final two considerations are the most novel. First, Collingridge and Reeve argue that it is characteristic of policy-oriented science that it leads to interminable disagreement. Rather than existing expertise being sufficient to determine policy or for additional research to be sufficient to narrow down the uncertainties, they suggest that, in policy contexts, further research tends to exacerbate existing uncertainties. A policy decision will almost inevitably produce some losers as well as winners. Those disadvantaged by the policy are likely to sponsor research into the basis of the science used in making the regulatory decisions. Associations of coal producers will commission studies intended to throw doubt on projections about the grave impacts of greenhouse-gas-driven climate change. Those who are likely to suffer from novel regulations will try to find unexpected benefits arising from their established procedures. They will look for weaknesses in the methods utilised in the studies which indicated that they were the ones at fault. They will attempt to show that alternative procedures have unexamined costs or impose unanticipated hazards. The more research produced, the less likely there is to be agreement. In this sense, Collingridge and Reeve directly adopt the leading finding of the Empirical Programme of Relativism and the Strong Programme, that agreement in science results from people deciding not to contest any longer, rather than from the debate having arrived at a point at which no further disagreement is logically possible. In a major policy dispute which threatens the economic and commercial interests of major corporations there are few incentives for participants to stop arguing and many inducements to argue for as long as possible.

In their view, this conclusion has a further implication: indeed a rather unorthodox one. These authors argue against the 'mythic' ideal that scientific advisers should be impartial or indifferent – Collingridge and Reeve use the more strident term 'irrelevance' – in relation to the consequences of the policies they offer. What they term the mythical 'principle of irrelevance' underwrites the idea that 'science can be of use to policy without becoming besmirched if the barrier between scientists and the users of their results is maintained' and that, 'Science is just as powerful whether its results are confined to the academy or applied outside to pressing matters of the moment' (1986: 28). Their view, however, is that experts' indifference is likely to serve policy actors poorly since, as Collingridge and Reeve assert:

> The principle of irrelevance, stating that the assessment of a scientific conjecture should be independent of any use to which the conjecture may be put, seems innocent enough at first, but on further analysis it must go, indeed it must be replaced by its converse, the principle of relevance, which holds that the uses to which any scientific conjecture is to be put shall always influence its assessment. This sounds quite shocking [at first]. (1986: 22)

Their point is that impartiality in pure science is based on a form of harmlessness. If one is disputing whether dinosaur extinction was caused by a colossal meteorite impact or by some much slower ecological changes, the stakes are high in scientific terms but their practical import is minimal. The dinosaurs are already doomed and little hangs (for now) on the answer. In basic science, researchers have time to consider and reconsider their ideas and they can happily entertain the abstract possibility that either hypothesis might be correct. By contrast, in policy work high stakes are likely to attach to the alternative courses of action. Accordingly, an indifference to the outcome is not politically feasible or morally defensible. Rather, policy advisers have to pay a great deal of attention to the costs of being wrong. This leads science for policy purposes to properly be more timid – more conservative in a certain sense – than basic science. Moreover, it may even mean that interpretations that are thought less likely to be correct but which have low costs may *quite reasonably* be initially preferred to ones that are thought more likely to be correct but whose consequential risks are higher. These analysts' (joyously unorthodox) advice is that policies should be adopted which are as insensitive as possible to advice arising from any scientific results (1986: 27).

Collingridge and Reeve summarise their argument for the over-critical model of science, advising as follows:

> On this model, whenever science attempts to influence policy, three necessary conditions for efficient scientific research and analysis – autonomy, disciplinarity and a low level of criticism – are immediately broken, leading to endless technical debate rather than the hoped-for consensus which can limit arguments about policy. The technical debate concerns the interpretation to be given to the existing body of evidence, but no matter how large this body may be, widely divergent interpretations may be maintained, making argument practically endless. As debate continues, many long-settled technical issues are reopened for investigation, and attempts to definitely resolve one issue often succeed only in opening up many more technical issues for consideration; technical uncertainties grow rather than diminish as more research is done. Relevance to policy increases the level of criticism to which technical conjectures are submitted, and such criticism is even easier than usual since the loss of autonomy and the weakening of disciplinary boundaries produces [*sic*] research results of poor quality. (1986: 145)

In other words, those features of scientific expertise which supposedly made it so suitable for advice-giving do not hold true in practice. And the actual features of science-for-policy make it far less than ideal as an aid to policy-making. Sound policies should be as independent of scientific advice as possible.

CLIMATE CHANGE AND THE OVER-CRITICAL MODEL: A CASE STUDY

The value of Collingridge and Reeve's analysis (1986) can usefully be gauged by using a case study that is a little different to the ones they chose. Still, one has to be careful to select a case in which there is a chance that scientific advising could follow the favourable pattern which these authors describe as 'mythic' and which they are aiming to discredit. In this context, it is fortunate that analysts of international relations and diplomacy have lately become interested in what are known as 'epistemic communities', cross-national communities of policy-advisers bound together by expertise. Recent authors have proposed that so-called epistemic communities are very influential in proposing, negotiating and implementing international agreements on such issues as international environmental policy. According to Haas:

> An epistemic community is a network of professionals with recognized expertise and competence in a particular domain and an authoritative claim to policy-relevant knowledge within that domain or issue-area.... [W]hat bonds members of an epistemic community is their shared belief or faith in the verity and the applicability of particular forms of knowledge or specific truths. (1992: 3)

The key assertion of epistemic-community authors is that members of such communities are able to come to agreed analyses of issues or problems with a degree of independence from their political bosses. Among biologists or atmospheric chemistry experts from different countries there exist 'intersubjective understandings' (Haas, 1992: 3). Accordingly, these expert communities' control over knowledge and information grants them independent power in shaping and co-ordinating international agreements. In other words, epistemic-community authors argue that in international relations something like Collingridge and Reeve's 'mythic' assumptions about scientific advisers do hold (for critical analysis see also Jasanoff, 1996).

An important case for gaining an insight into these matters of international scientific advising is that of climate change. For approximately the last two decades there have been attempts to lessen global warming by advocating curbs on the emission of greenhouse gases, notably carbon dioxide (CO_2). Scientists from many Northern nations who had been working on the global climate were brought together into a more formal arrangement in 1988 when the Intergovernmental Panel on Climate Change (IPCC) first convened (Boehmer-Christiansen, 1994a: 147). The IPCC office is based in Geneva and the work of the Panel was supported from the outset by the United Nations Environment Programme and by the World Meteorological Organization. The IPCC advised that states needed to act rapidly if greenhouse gas concentrations were to be regulated and it became clear that states had to respond in consort if success was to be achieved. Pressure grew for the introduction of some form of international treaty and in 1990 the United Nations took the initiative in setting up an Intergovernmental Negotiating Committee (INC) for a Framework Convention on Climate Change (FCCC) (Bodansky, 1994: 60).[2] The FCCC, set up in 1992, eventually gave rise to the Kyoto Protocol of 1997 which set out a process for introducing a binding treaty committing participating nations to greenhouse gas emission targets.

Most histories of this process have been written from the perspective of supporters of the FCCC. Even Boehmer-Christiansen (less of an enthusiast than most commentators) notes that, 'While by no means the first to involve scientists in an advisory role at the international level, the IPCC process has been the most extensive and influential effort so far' (1994b: 195). But far from science succeeding in resolving international environmental issues because of 'intersubjective understandings', existing suspicions of the industrialised North's interpretation of the world's environmental priorities have, in a number of cases, been exacerbated by the use of the supposedly impartial methods of science to diagnose the globe's problems. Thus in 1990, in an attempt to push policy-makers into acting on the findings of the IPCC and stimulating the setting up of the FCCC, the World Resources Institute (WRI) – a prestigious Washington-based think-tank – tried to set out figures indicating each country's CO_2 emissions in a year and thus their respective contributions to global warming (Dowie, 1995: 119). The WRI had shown an early interest in climate change and had been especially influential in publicising actual emissions-reduction targets against which governments' policies could be assessed (Pearce, 1991: 283–7). Their next task was to provide data on each country's carbon dioxide and other greenhouse-forcing pollution, allowing the appropriate amounts of 'blame' to be attached (World Resources Institute, 1990: 345). They chose 1987 emissions as their reference year; subsequently the FCCC has used 1990 as its point of reference.

Such a task faced many practical difficulties.[3] As even the IPCC had found, the data were hard to come by and countries had good reasons for concealing the extent of their emissions. There are many greenhouse gases, and their effects needed somehow to be integrated into a single greenhouse scale. But, in principle, the task seemed straightforward. From the point of view of global warming, one molecule of CO_2 is scientifically speaking the same as any other, so that different countries' emissions just need to be totted up and compared. According to the WRI's analysis, three of the top six net emitting nations were underdeveloped countries. In descending order the countries were the USA, the Soviet Union (still in existence then), Brazil, China, India and Japan. If all the EU member states were counted as one country it was even possible to argue that the remaining four places in the 'top ten' were occupied by the EU, Indonesia, Canada and Mexico. On this (possibly peculiar) view, fully half of the ten leading net emitters were from the non-industrialised world. This view came to win some acceptance in the scientific and policy communities and was repeated in mainstream texts (for instance in a chart in Pickering and Owen, 1994: 81).

While the WRI authors described their method as straightforward (see Hammond et al., 1991: 12), the study stimulated a fierce attack from Indian researchers based at the Centre for Science and Environment (CSE) in New Delhi. The CSE authors (Agarwal and Narain, 1991) offered several arguments in their critique. First off, they suggested that the sources for the figures were defective. For example, Agarwal and Narain presented evidence to suggest that the rate of rainforest clearance in Brazil was anomalously high in 1987 and that the felling rate declined considerably in the following year, due to changing financial incentives (1991: 4). Accordingly, the selected year gave Brazil a much higher apparent 'average' figure for CO_2 emissions (from burning the forest) then was truly average for recent years. Similarly they argued that the figures used to represent the loss of forest in India were based on old data from nearly a decade before when forest clearances were more common.

Though these problems of data gathering tended to cast developing countries in an unfavourable light by seeming to exaggerate their emissions, this was not the main focus for the CSE critique. Their more decisive argument was that the scientific language and apparent objectivity of the report's list of nations' respective contributions concealed two issues. The first had to do with assumptions implicit in the way 'net emissions' were calculated, while the second concerned the classification of types of emission. With regard to the latter, Agarwal and Narain suggested that it was unfair to compare CO_2 emissions which resulted from 'necessary' or unavoidable emissions – such as breathing – with ones that arose from entirely avoidable causes – such as driving to the supermarket to shop when one could have taken public transport. Surely these two forms of 'pollution' could not be equated; to count them all in together was to confuse different phenomena.

Their first point is more complex: like any modelling exercise, the procedures used for the WRI report depended on certain assumptions. A key assumption concerned 'sinks' for the greenhouse gases. When carbon dioxide is released into the atmosphere, not all of it remains in the gaseous state in which it performs its greenhouse 'function'. Some carbon dioxide becomes dissolved by rain or is directly absorbed into the oceans, while a large amount is taken up by plants and soils. Indeed, the annual natural cycling of carbon through the atmosphere greatly exceeds the amounts added by human activity (Pickering and Owen, 1994: 83). Less than half the additional carbon pushed into the atmosphere each year by human activities remains there; according to both the WRI and CSE, over 56 per cent of humanly produced carbon dioxide is absorbed by environmental sinks. The figures for other greenhouse gases, notably methane at around 83 per cent, can be even higher.

The methodology of the WRI study recognised this fact. Loosely speaking, it was taken into account by discounting all emissions by the rate at which they are absorbed. In other words, if 56 per cent of all CO_2 emitted each year is re-absorbed by trees, marine creatures and the oceans, countries are actually only causing a warming in proportion to 44 per cent of the total amount of carbon dioxide they emit. On the face of it this seems perfectly reasonable. However, Agarwal and Narain contended that this was actually unfair because it shared out the natural sinks in proportion to how big a polluter each country was, since one received a 'discount' for every single molecule of pollutant emitted. They advocated an alternative approach which treated the sinks as something like the natural patrimony of the whole human race. On this approach, one might wish to add up the absorptive capacity of all the natural sinks and then divide them equally between the global population. One could then allocate 'shares' to the various countries according to the size of their populations; only at this point would each nation's emissions be reduced by the appropriate discount. Although India as a nation is a large greenhouse polluter, this is because it has a very large population with nearly four times as many people as the USA. On the WRI figures, the average Indian citizen has a greenhouse impact of roughly one tenth of that of Britons and Germans and even less when compared to the average US citizen. However, if one treats the figures in the CSE's fashion, it appears that Chinese and Indian people are actually living within the limits of the natural cycling capacity of the planet, whereas North Americans, the Japanese and Europeans are not.

To express this another way, if everyone on the planet only emitted at Indian or Chinese per capita levels, the natural sinks could be expected to cope easily with all the greenhouse pollution from carbon dioxide. The same reasoning applies to methane as well. As Agarwal and Narain put it:

> WRI's legerdemain actually lies in the manner that the earth's ability to clean up the two greenhouse gases of carbon dioxide and methane – a global common of extreme importance – has been unfairly allocated to different countries. ... Global warming is caused by overexceeding [the] cleansing capacity of the earth's ecological systems. The WRI report makes no distinction between those countries which have eaten up this ecological capital by exceeding the world's

> absorptive capacity and those countries which have emitted gases well within the world's cleansing capacity. (1991: 10)

On this view, the average Indian citizen of 1987, when allocated their 'share' of the world's carbon sinks, is actually a net absorber of carbon dioxide. This conclusion seems to bear out Collingridge and Reeve's claims very strongly since it fits their suggestion that 'attempts to definitely resolve one issue often succeed only in opening up many more technical issues for consideration' (1986: 145). The attempt by the WRI to draw up a definitive emissions 'league table' succeeded only in opening up international disputes about the measurement and conceptualisation of carbon sinks. Interestingly, in this case, one of the issues opened up is explicitly ethical as well as technical.

Ironically, in many respects Agarwal and Narain seem every bit as attached to the myths of policy-making as the WRI. In a notably unreflective move, these authors treat it as more or less unproblematic to assign 'shares' in the global sinks in proportion to various countries' populations, without observing that this approach is not self-evidently correct either. Such a method, though straightforwardly related at some level to principles of equity, has the drawback that it effectively rewards countries for increasing their populations. Furthermore it allocates sinks while taking the 'nation' entirely for granted as the unit of analysis. Subsequent developments in the Soviet Union indicated what fragile constructions 'nations' may be. Moreover, Agarwal and Narain pay no attention to the allocation of resources between, say, rich and poor within countries or between women and men. One could easily imagine a feminist critique which added up men's and women's emissions and implied that men are the greater carbon villains.

Still, it should be pointed out that Agarwal and Narain's concern with sinks is by no means quixotic. By the time of the Kyoto Protocol, this focus on the treatment of sinks had made it onto the official agenda – in part to assist some countries in complying with targets. According to the Protocol, countries are able to include 'carbon "sinks", that is, emissions "sequestered" by assorted land use changes, such as the planting of new forests or forgoing planned land clearance' (Boehmer-Christiansen, 2003: 71). On this view, a country's entitlements to sinks is not derived by adding up all the world's sinks and allocating it to nations per head of population. Nations can manage their own sinks (for example by planting additional forests) and retain all the CO_2-reduction credits themselves; countries can even cut their calculated carbon emissions to some extent by funding the development of sinks in other countries. Of course, this is to take a different view of sinks from that proposed by Agarwal and Narain; sinks are now appropriated by individual nations while Agarwal and Narain took them to be part of humanity's common inheritance.[4]

LESSONS FROM THE CLIMATE CHANGE CASE: HOW TO UNDERSTAND SCIENTISTS AS POLICY-ADVISERS

Despite the apparent suitability of climate change policy as an issue to be directed by an epistemic community, this case study has indicated that even in this instance the policy process has instead had many of the characteristics anticipated by Collingridge and Reeve. According to the epistemic communities view, the scientists within these specialised groupings would be expected to come to agreement much more readily than do nations' political representatives and thus the epistemic communities generate leadership and influence on the strength of their professional agreements. However, the climate change case exhibits many features which fit uneasily with this view. Even the apparently simple, empirical matter of working out how much greenhouse gas each country should be held responsible for turned out to be complex and contested. Just as Collingridge and Reeve (1986) had proposed, the high stakes led researchers allied to different interests to try to deconstruct the unwelcome conclusions with which they were confronted by other scientists. Where the WRI claimed to see a simple counting exercise, Agarwal and Narain detected environmental colonialism. Research on climate issues did not lead people to agree; instead it often reinforced their confidence in their differences. Neither did research on the key issue of sinks lead to unanimity. Instead, it turned out that sinks can be conceptualised in a variety of ways; the ways people favour seem to vary from one political standpoint to another.

The climate change case exhibits other features identified by Collingridge and Reeve also. Thus, much publicity has lately been given to the proposal that global warming has resulted from (possibly cyclical) variations in the heat energy released from the sun as much as from atmospheric changes. In this instance, as Boehmer-Christiansen notes, knowledge claims from other disciplinary sources can be called on to question the climate-science community's view: 'space physics in general with NASA and the European Space Association in the lead are now the main challengers to [the] IPCC by testing the role of clouds, cosmic rays and solar phenomena' in climate change (2003: 77). This dispute exemplifies Collingridge and Reeve's point about disciplinary disagreements: most atmospheric modellers exclude the science of the sun from their models; they commonly hold solar emissions constant. For their part, when assessing changing temperatures on the earth, solar scientists tend to grant explanatory priority to alterations in the sun and the heat it radiates. Finally, the issue of humanly caused climate change is a focus of such intense scientific scrutiny not because it is a topic autonomously selected by the scientific community. Rather, the scope for intervention in the growing policy field of climate 'management' created the opening for a research endeavour focused on the IPCC. The 'problem' created the field, rather than the field discovering the problem.

However, this does not imply that Collingridge and Reeve are entirely correct. Their third and distinctive assertion (that further research tends to exacerbate existing uncertainties in policy disputes) is an insightful development of the central claim of science studies – namely, that agreements about the natural world are achieved by people and not by the dictates of the natural world. But, as was made clear in Chapter 10, certain institutions seem to be better 'designed' for undermining scientific expertise, and for discouraging agreement, than others. It appears, for example, that legal cross-examination is particularly effective in this regard. Seeing the potential for endless deconstruction which may result from more and more research, Collingridge and Reeve propose that the best policy options are the least research-sensitive ones. They identify themselves as pragmatists, as incrementalists. But this results in their position almost becoming 'research-averse'. Their desire to attack mainstream notions of the role of science-advising in the policy process leads them to place all their emphasis on the reasons that competing expert policy-advisers may have for disagreeing. As a result, they pay little attention to the ways in which agreement may be fostered. Even if one accepts that agreements about the natural world (about global warming for example) are achieved only by people, this does not mean that one should not wish to stimulate agreements which are openly arrived at. Only if one does not believe in the possibility of agreement at all would it be sensible to adopt a comprehensively research-averse position.

An alternative conclusion would be to favour a move away from the predominant models of expert advice-giving (for a helpful step in this direction see Turner, 2001). This would involve paying more attention to the institutions of policy-advising and trying to steer these clear of quasi-legal contests and other formats which inhibit agreement. As Jasanoff and Wynne have recently commented, social studies of science is ideally suited for this analytical task:

> Constructivist approaches in the social sciences illuminate the extent to which our knowledge of the global environment is made by human agency, and not simply given to us by nature. In particular, interpretive analyses of the framing of policy problems, the production of scientific claims, the standardization of science and technology, and the international diffusion of facts and artifacts all focus attention on the co-production of natural and social order. ... Thus, they also provide a more textured and useful account of how scientific knowledge becomes (or fails to become) robust in policy contexts. (1998: 74)

Despite the strength of their analyses, Collingridge and Reeve have in a sense a too intellectualised notion of science for policy. The key result of a science-studies informed approach to the question of science's role in policy-advising is not that sensible policy is research averse but that analysts need to study the building of policy advice and policy-advising institutions at the same time. In the next chapter a different kind of limit to the advice-giving ability of science – a more inherent form of limitation – will be considered.

---

[1]For a recent version of this claim see the latest UK official statement of its rationale for science-funding (UK Government, 1993).
[2]Loosely defined, a 'Framework Convention' is an undertaking to set up a forum committed to certain objectives within which particular binding agreements will subsequently be developed in the form of 'Protocols'. Thus the FCCC contained no specific greenhouse-gas abatement undertakings, only an agreement to develop and possibly engage in such arrangements in future years; it did however set out certain procedural matters concerning decision-making and so on. Many international agreements take this form.
[3]The following account draws heavily on the fuller analysis in Yearley, 1996: 100–21.
[4]Of course, challenges to the IPCC and associated agencies came from other political quarters also. In 2001, the US President (George W. Bush) declined to even try to get the US legislature to ratify the Kyoto Protocol, citing scientific uncertainty both about the causes of climate change and about the consequences of continued emissions, and claiming that the economic costs of compliance with the treaty would be too severe for the US economy (see McCright and Dunlap, 2000). Other critics of the Kyoto Protocol emphasise that compliance with the agreement will be very expensive for all industrialised countries but will result over the next decades in only marginally lower temperature increases than would have occurred in any event since, with or without Kyoto, the overall concentration of greenhouse gases in the atmosphere will still increase (see Boehmer-Christiansen, 2003: 70 and 89).

12 Conclusion: Science Studies and the 'Crisis' of Representation

DIAGNOSING THE CULTURAL CRISIS OF SCIENTIFIC AUTHORITY

As we have seen, at the outset of the twenty-first century the scientific profession faces significant problems in relation to its public standing and credibility. There is reluctance to accept experts' calculations of risk, even an unwillingness to accept the entire cost-benefit framework within which risk is conceptualised. The scientific establishment itself acknowledges and is trying to counter difficulties in what it sees as the public's (mis)understanding of science. Problems beset attempts to work out which scientific evidence to admit into court. And, in the arena of science advising, it seems impossible to realise the ideal that scientists should speak truth to power; severe complications confound officials' relationship to providers of expert advice. It seems that, in every quarter, scientific representations of the way that the world is, are being questioned or rejected or discounted. Some commentators (most explicitly Redner, 1994) suggest that this state of affairs is best interpreted as part of a more widespread challenge threatening the ideal of representation in contemporary culture. In the face of a series of challenges to accepted notions of representation in different intellectual and creative fields, such commentators suggest that it makes sense to try to find an underlying explanation for the simultaneous appearance of these problems. How then should this malaise be understood? One frequently heard and apparently comprehensive answer is that this fix in which scientific representation finds itself is part of a more general cultural crisis, usually described by the term postmodernism. Seemingly in every cultural sphere – in art, in literature, in design – the idea of a definitively superior style and the ideas of advance and progress have come to be questioned. These days, plurality and multiplicity are preferred to adherence to the one privileged representational technique. Maybe the difficulties besetting scientific expertise are a sign that this is happening in science too.

To assess this line of reasoning I need briefly to outline what I take to be the relevant aspects of postmodernists' claims. Postmodernists argue that, until recently, intellectual and artistic endeavours were spurred by a modern (or perhaps one should say modernist) belief in progress. Representation – whether that was artistic depiction or scientific description and modelling – was the goal, and debates centred on the best means of accomplishing representation. In this way successive innovations in artistic representation challenged the presumption of the former style to be properly or adequately realistic. Early modern artists introduced the conventions of 'true' perspective into painting and may have used pinhole-camera-like devices to generate perspectival images. The handling of light and shade, of the gradients of colour, became more self-conscious and sophisticated. In the nineteenth century photographs were used to show what things really looked like frozen in a single moment of time in order to support 'improvements' in artists' depictions of animals or cloud patterns and sunlight or the weather. Also in the nineteenth century, the Impressionists had challenged their predecessors' claims to realistic depiction by asserting that their own representations were truer to the momentary reality of perception. But in the twentieth century this progressive trajectory ran into the buffers.

The essence of the postmodern analysis of art was (complexly) summed up by Lyotard in his essay 'What is postmodernism?' as follows:

> What space does Cézanne challenge? The Impressionists'. What object do Picasso and Braque attack? Cézanne's. What presupposition does Duchamp break with in 1912? That which says one must make a painting, be it cubist. And [Daniel] Buren questions that other presupposition which he believes had survived untouched by the work of Duchamp: the place of presentation of the work. In an amazing acceleration, the generations precipitate themselves. (1984: 79)

The key sentence here (and sadly the most obscure) is the last. The point is that, with gathering speed, the innovators outbid each other and, in the process, appear to undo the very rules of the contest in which they are engaged. The Cubists present objects as seen from multiple perspectives. Duchamp introduces the 'ready-made' and displays not a still-life depicting flowers and jugs on a table-cloth but an actual, shop-bought urinal and a pot rack. But at least he displayed such pieces in museums and galleries as works of art. Buren, the conceptual artist, rationalises still further and leads his art outside the gallery. His installations spread outwards from the gallery space around the exterior of the building; at other times he put the work on billboards and had it carried around by gents with sandwich boards. He breaches the gallery/non-art-world divide.

So, loosely expressed, the suggestion is that artists of all kinds used to do whatever it was they did straightforwardly. The modern movement put paid to that by insisting that artists reflect on and rationally improve what they do. The history of art acquired a directionality. But then this modernist impulse to be self-critical about what you are doing became self-destructive. Modernism enters a crisis of its own creation and the post-modern condition arises when people realise that the modernist game is up. The race for better representation cannot be won; as one seems to near one's goal, the goal itself comes to appear illusory and ludicrous. Duchamp's 'ready-mades' are immensely realistic just because they are the real thing. It looks like he holds the winning cards but his starkly triumphant hand actually marks the end of the game.

The point is not simply that the artistic community turns away from modernist objectives and elects to do something else instead; it is not that modernist objectives are seen as possible but undesirable. These objectives are actually undermined from within; they are laid to rest and transcended. Postmodernism is liberating because it removes the old ambition as well as the old constraints. Artists no longer strive for the one best way of telling narratives, of designing buildings or of painting portraits. The very idea of a definitive representation or of the 'best' design is rejected and in its place a welcome is extended to a plurality of versions which offer a multiple rendering of any subject.

FINDING POSTMODERNITY WITHIN SCIENCE

The postmodern condition in art, say, or in architecture leads to a loss of authority by the Academy. Without agreement about the objectives and about the appropriate means for doing artistic representation, it becomes much harder to regulate who is and who is not an artist or leading architect. The analogy with the predicament of science is appealing. Who is to say that Greenpeace is not an expert on the risks of biotechnology or that alternative therapists' views on MMR should not be taken as an authoritative statement on the sources of well-being. Even if one presses the analogy a little harder it appears to hold. Both science and art are concerned with representation. Both areas have witnessed disputes over the nature of proper representation: every recent art prize competition seems to generate well-publicised disagreements about the status of the works as art while, as mentioned above, there has been comparable media attention paid to public disquiet over the legitimacy of scientists' accounts of the risks that the public faces from genetically modified foods or from the nuclear industry. In both cases there is a babble of claims and counter-claims. The symptoms seem to fit; scientific authority is in a postmodern fix.

Authors such as Lyotard have been quick to advance this argument, drawing not just on the similarities in the public unease and ensuing public disputes over scientific and artistic representation, but on features of science itself. They claim to find evidence in the development of scientific ideas for its postmodern condition. By and large, authors writing on this theme offer the same list of trophies from their hunt through contemporary science: quantum uncertainty, chaos theory and catastrophe theory.

Taking these briefly in turn, within quantum physics these commentators focus on the notion that the state of the physical world can be affected by the very action of observing it. They are pleased to find that science now teaches that one cannot observe quantum reality without influencing the reality under observation. The old idea of objective observation and representation is thus subtly undermined. Accordingly, Lyotard argues that in quantum theory and microphysics, 'The quest for precision is limited ... by the very nature of matter' (1984: 56). Similarly, Redner invokes Heisenberg's uncertainty principle to illustrate the 'interaction between observer and "object"' (1987: 68). The next candidate, chaos theory, is presented as the scientific community's way of understanding systems which appear too complex to submit to conventional scientific models. Thus, weather prediction for more than 10 or 11 days ahead continues to confound scientists' efforts. Until comparatively recently this was held to be because existing computational models of the weather could not adequately represent the way in which the multiple influential factors impacted on each other. It was assumed that better computing could address such problems. Chaos theory suggests that it is not so much the multiplicity of factors that is important; rather it is the inherent character of the relationship between those factors which makes the weather unpredictable in principle. For Redner, chaos is important as a challenge to the assumptions of classical science. Lyotard too invokes ideas around chaos to support the idea that there is 'a current in contemporary mathematics that questions the very possibility of precise measurement and thus the prediction of the behaviour of objects even on the human scale' (1984: 58). Finally, there is catastrophe theory which offers a mathematical approach to understanding why stable systems can break down suddenly, why structures collapse without warning or, in Lyotard's favoured example, why enraged but fearful dogs suddenly bite (1984: 59).[1] Redner describes it as 'a theory of sudden, discontinuous movements, where small changes can generate large effects – the so-called catastrophes' (1987: 276).

So, it would seem, the game is up for science – that archetypal modernist activity – too. Lyotard suggests that there is a 'crisis of determinism' (1984: 53). At the most fundamental, microphysical level, the natural world resists unambiguous scientific representation. And even at the level of macroscopic (even planetary) phenomena – such as weather forecasts two weeks into the future – there is no prospect of drawing definitive representations. Surely, we are invited to think, this inability to produce compelling representations of how the natural world is must lie at the heart of the credibility and legitimacy problems confronting science. Postmodernists seem to offer a theory which comprehensively accounts for the decline in the authority of science.

EVALUATING THE POSTMODERN DIAGNOSIS

Postmodernism has enjoyed a high degree of popularity in the social sciences but, against the views of many commentators, I wish to argue that the postmodern diagnosis of the malaise of scientific authority is wrong on two counts. First, the areas in which the credibility of science is most conspicuously under pressure are not, by and large, the ones which feature on the postmodernists' problem page. The areas of science involved are commonly not the ones whose problems most easily fit with the supposed 'crisis of determinism'. As we have seen, the credibility of scientific expertise has been challenged over the ability to estimate risk from nuclear installations or biotechnology, over the reliability of expert testimony in court and over the openness of the medical establishment in acknowledging uncertainty concerning the safety of treatments or the viability of alternative remedies. In the artistic case, the issues provoking public unease were also the ones which excited postmodernists: for instance the status of conceptual 'art' as art. For the case of science this is much less evidently the state of affairs. Admittedly, scientists have been challenged over the credibility of their advice on global warming, a topic which might be seen as allied to some degree to the matter of the weather's unpredictability. But even here the difficulties are as much ones of conventional science – of supposed entrenched interests and the comprehensiveness of models – as they are of post-deterministic science.

Conversely, the 'trophy' elements of science paraded by postmodernists are not themselves the ones provoking public scepticism. The misleading nature of the attempt to detect the postmodern problem of credible legitimation in science itself can be illustrated with regard to chaos theory. The central claim here is that, within the last 30 years, it has been realised that there may be something in common among many natural phenomena that appear to be complex and unpredictable. The exact shapes of coastlines for example do not seem to correspond to any recognised geometrical or standard mathematical patterns. It has long been accepted that some things may be complex because they are dependent on very large numbers of variables and thus, in practical terms, are beyond calculation. But authors interested in chaos are particularly fascinated by the possibility of phenomena which appear complex and unpredictable but which none the less are based on relatively simple relationships.

The distinctive idea is that of 'deterministic chaos': the suggestion that apparently straightforward equations can give rise to chaotic, seemingly random trends. Hitherto, work in applied mathematics tended to focus on those equations that underlie orderly and simple behaviour. These are the equations that describe lines or circles, the elliptical movement of the planets or the parabolic paths of comets and so on. But other equations, not necessarily more complex in appearance, generate different sorts of results. The trends which they describe are not at all geometrically regular. With a straight line or regular curve, one can readily predict the next point from the foregoing ones, even without calculating it. But in the case of irregular equations no such prediction is possible since the points veer around in an unforeseeable way. The former sort of equation is commonly referred to as linear while the latter one is, unsurprisingly, called non-linear.[2]

In some cases these non-linear equations describe shapes which are not geometrical in the standard sense but which are none the less recognisable. They don't look like triangles or circles but they may resemble the shapes of complicated leaves or of the coastline seen from high altitude or of the apparently irregular peaks of mountain ranges. In his celebrated play *Arcadia*, Stoppard introduces a fictional early innovator in this area of study, a precocious teenager in the early nineteenth century, called Thomasina. She argues with her tutor, Septimus Hodge, about the adequacy of the (linear) mathematics he is teaching her for understanding the variety of forms found in nature:

> THOMASINA: Each week I plot your equations dot for dot, *x*s against *y*s in all manner of algebraical relation, and every week they draw themselves as commonplace geometry, as if the world of forms were nothing but arcs and angles. God's truth, Septimus, if there is an equation for a curve like a bell, there must be an equation for one like a bluebell, and if a bluebell, why not a rose? Do we believe nature is written in numbers?
>
> SEPTIMUS: We do.
>
> THOMASINA: Then why do your equations only describe the shapes of manufacture?
>
> SEPTIMUS: I do not know.
>
> THOMASINA: Armed thus, God could only make a cabinet.
>
> (Act 1, Scene 3 in Stoppard, 1999: 55 – see also 118)

The mathematics she is learning is able to describe simple geometrical shapes, such as those used to fashion square tables and round stools, as well as those employed to describe the orbits of the planets. But this mathematics disappoints her by failing to account for the shape of a rabbit's ears or of a pine tree. Exasperated, she complains that – restricted to these shapes – the creator could only have made trees and hedges in the shape of rectangular wardrobes or conical hats.

Stoppard allows Thomasina, and subsequently Septimus, to become precursors of a new mathematics, using non-linear equations to produce curves in the shapes of leaves, of trees and of animals' features. Stoppard is able to play this anachronistic game precisely because some non-linear equations are relatively simple. The calculations are just extremely laborious, a problem overcome in the play when lovesick Septimus adopts a hermit's lifestyle, affording him ample time to work out his sums.

The key implications of deterministic chaos flow from problems arising when one tries to predict the future state of non-linear, chaotic systems. With all equations (linear or non-linear) the exactness of one's prediction of any trend depends on how precisely one understands the starting conditions. The collision between the cue ball and one of the target balls on a pool table is often held up as an example of (linear) Newtonian physics but, even in this case, the precision with which one can predict the future disposition of the balls is believed to depend (along with other incidental things such as the friction of the cloth and the bounciness of the 'cushes') on knowing precisely how fast and in precisely which direction the white ball is moving in the first place. Small or larger errors in this knowledge will lead to small or larger errors in prediction and, the further in time one goes on, the greater these inaccuracies are likely to become. Non-linear equations share this dependence on knowing the starting conditions. But since the behaviour which they underlie does not follow a predictable trend, minor inaccuracies in the starting conditions can lead to radical errors in prediction. Errors in gauging the initial conditions of a pool-table collision may lead to a ball missing the hole or some other kind of foul shot. But in a non-linear system, the error might not be this minor. The ball might, so to speak, end up on the floor under the table or in the car park outside the pool hall. Or, as Series and Davies put it, 'any input error multiplies itself at an escalating rate as a function of prediction time, so that before long it engulfs the calculation, and all predictive power is lost. Small input errors thus swell to calculation-wrecking size in very short order' (cited in Carey, 1995: 500). This produces an apparent paradox: behaviour which is deterministic – which, that is, is believed to be governed by an algebraical equation – can be effectively unpredictable.

Hitherto, the unpredictability of processes that appeared to defy prediction was generally interpreted as arising from the interplay of numerous, incalculably interacting factors. For example, in roulette we have devised a thoroughly unpredictable system; this is after all what makes it ideal for gambling. In this case the unpredictability arises because the light ball and the heavy turntable with a ridged and curving surface, spun at slightly different speeds on successive occasions, combine to make the outcome in practice impossible to calculate. With deterministic chaos, however, a similar degree of unpredictability arises from the working-out of a relatively simple equation.

The existence of these views in the scientific community about such equations has two complementary, though contrasting, implications for our assessment of the postmodernists' verdict. First, it suggests that the scientific community is less disturbed than Redner and Lyotard would have us believe by the notion that processes may be governed by scientific laws but none the less still defy prediction. As is widely known, it is commonly suggested that the weather system follows this pattern. The movements of air currents, the condensation of moisture and so on are all deterministic, but they are also chaotic. Accordingly it is impossible to predict the weather more than around 10 or 11 days ahead since, by that time, inevitable errors in assessing the starting conditions will have magnified to the same scale as the meteorological system itself. The contemporary cliché about the weather in Tokyo being affected by a butterfly's wing beats in the Amazon is a 'poetic' expression of this point. An error in specifying the initial conditions – even one as slight as overlooking the minor changes in air patterns caused by the beat of a butterfly's wings – could, after little more than a week, develop into huge inaccuracies in projections about the future state of the atmosphere. Understandably, such sensitivity to initial conditions is believed to impose an inherent limit on the potential for weather forecasting (see Palmer, 1992). This phenomenon of acute sensitivity to the initial condition of the system is believed to be widespread. Such sensitivity is claimed for population patterns also, making, for example, the modelling of fish populations or of the growth of 'plagues' of insects, at least under certain circumstances, equally intractable. Simmons, a geographical commentator, asserts that 'This throws into doubt any claims that ecology might have had as an over-arching narrative for the whole of human–environment relations: like postmodernism in social and literary theory, the foundations are being questioned' (1993: 35). But, in his haste to make connections between the natural sciences and other aspects of culture, Simmons seems to be arguing with less care than usual. One can acknowledge that the ubiquity of non-linear systems does seem to make advances in scientific understanding appear significantly more modest than they had been thought to be, since major, consequential aspects of the world appear to be inherently beyond scientific prediction. But the irony is that they defy prediction for wholly scientific reasons. Ideas about chaos are not the vindication of postmodernity but something rather more oblique as the next point illustrates.

The second point is that the suggestion that apparently chaotic behaviour may be underlain by deterministic formulae can, ironically, be used to legitimate scientists' attempts to extend the reach of scientific representations. Scientists can claim that phenomena which previously appeared simply random may instead be viewed as deterministic and amenable to understanding in terms of equations. This argument has been advanced in relation to plant and animal population figures as Simmons indicates (1993: 35). These seemed to fluctuate wildly, even randomly, and thus were often taken to be beyond scientific understanding. Since approximately the 1970s, biologists have argued differently, claiming that population trends – which often swing from abundance to scarcity and back again with no apparent logic or regularity – are in fact following mathematical laws. Robert May, one of the innovators in this field, proposed that other phenomena such as animals' colour markings and certain movements of the market – things which appear truly random – may also turn out to be governed by straightforward equations (see May, 1992). On this view, chaos represents not the end of science's ambitions but a new extension for scientific understanding since we now 'realize that extraordinarily complex behaviour can be generated by the simplest of rules' (May, cited in Carey, 1995: 504). Discovering a seemingly random pattern in nature, the scientific analyst may now assume not that there are myriad complex causes at work but surmise that there is an underlying pattern, even if the rule which describes that pattern has not as yet been figured out. Hence the key assumption of chaos theorists is that much of the world that is unpredictable and apparently complex is actually deterministically chaotic. It is believed to be underlain by simple equations but in laboriously complex relation to each other. In principle, one can calculate them, but the labour of calculation grows ever more sharply. Scientific commentators suggest that, in the long run, events will unfold more hurriedly than they could ever be calculated. Or, as Davies strikingly puts it, 'the Universe is its own fastest simulator' (cited in Carey, 1995: 501). On this view, the future is in principle unknowable, even if it is not exactly unpredictable. At first sight, science seems to be in a postmodern fix. Scientific reasoning spells out its own limitations. At the same time, deterministic chaos is a rallying cry for new interpretations, for the advance of scientific determinism into areas previously regarded as off-limits.

To summarise, therefore: the diagnosis that the credibility problems of science are evidence of, and arise from, the hazards of postmodernity is unconvincing for two reasons. The 'trophy' examples such as chaos do not work out the way that advocates of the postmodernist diagnosis would wish. Chaos does not amount to a 'crisis' for science in the way that conceptual art does for the art world. Non-linear equations are still said to be deterministic. Chaos is viewed as the specialised characteristic of particular phenomena in science, not a challenge to the whole idea of modelling the world mathematically. On the contrary, analyses in terms of chaos suggest that some phenomena formerly regarded as random are governed by mathematical formulae. Secondly, the areas where the public 'crisis' of representation is most conspicuous are not particularly the ones characterised by chaos. Rather, the crisis is at least as much about the confidence people can have in the scientific community to (for example) assess risks in a way people feel they can sympathise with, as it is to do with the inherent limitations of scientific prediction and understanding.

In the art case, at least as Lyotard tells it, it is the realisation of the folly of the modernist project that brings an end to modernism and creates the public crisis of the arts, since in post-representational art 'anything goes'. Conspicuously, the essence of Lyotard's argument seems to be that the crisis of postmodernism is an internal malaise – a kind of over-ripeness. Artists' reflection on the business of representation finally undermines the enterprise from the inside, and the crisis of science is presented in the same way too. Lyotard may have a point about art, though it could easily turn out that historians of art find this narrative unpersuasive; certainly the story he advances seems to be an almost exclusively intellectual one. In the case of science, such a story simply does not hold water.

AN ALTERNATIVE DIAGNOSIS

To reject the postmodernist account is not to deny that there is a widespread problem with many scientific representations of the natural world and with the credibility of science. There is, and the best understanding of the underlying grounds for this come not from postmodernism but from science studies.

As argued in Chapter 7, the key realisation is that scientific agreement typically arises from people consenting to stop disagreeing rather than from the compellingness of scientific evidence itself. There are always conceivable grounds for resisting scientific interpretations; agreement arises when no one is motivated or able to dissent any longer. Because the world which natural science sets out to represent is inevitably inscrutable, because – that is – there is no way to learn about the world than through science and no means to check scientific understandings than through other understandings, the scientific community has only ever had its own methods to certify that its representations are correct. The claims of science have to be checked using other scientific claims. For this reason, the business of representation has, as ethnomethodological studies of science reminded us in Chapter 6, always been simultaneously ordinary and precarious. In the ethnomethodological senses, the work *of* science is to build and defend representations. By contrast, the validity of artistic representation has typically been affirmed in more straightforward ways. The world portrayed by the artist has, until recently at least, been highly scrutable and subject to assessment by a wider public or group of consumers. In that peculiar sense one could just about claim that science has always operated approximately in the way in which postmodernists see culture operating from the mid/late twentieth century onwards. In principle, there is nothing new about this crisis of representation in science; it has been fraught with precisely these difficulties since the time of the scientific revolution.

Latour too is impatient with postmodernists' claims. He makes the droll argument that their concern about postmodernity is rather rich since we never even managed to become modern in the first place. The modern 'constitution', he suggests, depends on the idea that one can achieve in practice the mental segregation of humans from the natural world, divide culture from nature, and goal-directed action from mere behaviour. But this segregation has never been achieved and shows no signs of being attained today when so many hybrids, entities which are both cultural and natural (computer intelligences, people with pacemakers, apes schooled in human sign language[3]), indivisibly bind humans to the natural world. His diagnosis, also the title of his book (originally *Nous n'avons jamais été modernes*, 1993), is that 'we have never been modern'. In the light of the arguments made in the last paragraph, I suggest it would be more accurate to say, '*La science, elle a toujours été postmoderne*' (Science has always been postmodern).

In offering this slogan I do not mean to claim that scientific knowledge and its social role have forever been constant. As the scientific community has promoted itself as the source of answers to society's problems and as policy-makers have come to consider delegating decisions to scientific advisers, the problematic features of science as a form of knowledge have become exposed. The courts have found that one cannot specify rules for what should count as science; policy-makers have fallen foul of the over-critical model of science advising. Beck's idea of reflexive modernisation is closer to the mark than the diagnosis offered by Lyotard or Redner. But even Beck is too focused on philosophical ideals, on claims about over-ripeness:

> ... science becomes more and more *necessary*, but at the same time, *less and less sufficient* for the socially binding definition of truth. This loss of function is no accident. Nor is it imposed on the sciences from outside. ... On the one hand, as it encounters itself in both its internal and external relations, science begins to extend the methodological power of its skepticism to its own foundations and practical results. Accordingly, the claim to knowledge and enlightenment is systematically scaled back in the face of the *successfully* advanced fallibilism The access to reality and truth which was imputed to science at first is replaced by decisions, rules and conventions which could

> just as well have turned out differently. Demystification spreads to the demystifier and in so doing changes the conditions of demystification. (1992: 156, original emphases)

But the destructive work of demystification is not principally performed by science itself (nor by the philosophy or social studies of science which Beck would likely also classify as 'sciences' in this sense). It is performed in courts, in challenges to risk assessments, in disputes over public policy or in the public's rejection of the views of the scientific establishment (on the last of these see Wynne, 1996). The 'reality' about which science seeks to inform us has always been inscrutable; scientific knowledge has always been 'postmodern'. When it is primarily used for regulatory purposes and is therefore subject to relentless challenge its 'postmodern' frailties are revealed.

These factors go further towards explaining the predicament of science than the more abstract assertions of postmodernists. Ironically, these factors appear as problems today because of the high social profile of science. The scientific establishment's own successes have opened science to deconstruction and to scepticism.

CONCLUSIONS

This book has been about science studies and social theory. Its argument has been that sociologists and social theorists have paid too little attention to society's dark matter – to the role in society and social life of scientific evidence, of technical expertise, of scientific laws and of 'actors' such as risks and technological systems. When sociologists have sought to incorporate insights about such topics into their work they have paid unjustifiably little attention to analyses in social studies of science and taken rather too much notice of writers on postmodernism or on risk communication and so on. My aim in the first two Parts of the book was to promote sociology's interest in and appreciation of science studies by reviewing the state of the art in this field. In the third Part, I used a series of themes of greater familiarity to social science disciplines (such as the assessment of risk and the nature of legal judgement) to offer an indication of the analytical value of science-studies work in enhancing sociological understanding of the significance of science and technical expertise.

In addition to this 'evangelical' and expository intent, two major themes have pervaded this book and make up its analytical conclusions. The first of these was introduced at the outset as the issue of dark matter or the missing masses. Social life is facilitated, conducted and performed through the world of things. Science and technology are our principal means of interpreting and representing these things, yet scientific knowledge and scientific practices are routinely under-emphasised by almost all social theorists. Without an understanding of the dynamics of science, the sociological analyst will arrive at only a very partial account of society's operation. For example, the detailed analysis (in Chapter 9) of how risk assessments are put together, and how similarities and differences between various sorts of risks are constructed, showed that a science-studies approach can throw light on the operation of society's dark matter, more precise light indeed than other social scientific approaches to risk.

The second analytic theme concerns the idea that scientists' analyses of scientific theories and findings are autonomous and objective. On this view, science itself offers the best (and only) account of how society's missing masses function. In other words, sociologists could acknowledge that scientific objects are the missing masses of social science but still delegate the understanding of those objects exclusively to the natural sciences. That option has been explored and rejected in this book. The point of view I have argued for is that beliefs about the state of the natural world are ultimately decided by groups of people (usually scientists, but people none the less); such beliefs are not compelled or fully determined by information from the natural world itself. There is therefore an inescapable sociological dimension to the understanding of society's missing masses.

As the book has made plain, the various schools of thought in science studies diverge in their interpretations of how such decisions about the state of the natural world are taken and shaped. The third Part of the book has been developed as a means of carrying out a practical test of the value of those competing schools of thought. Key issues concerning risk, the public's understanding of science, science and policy, and science in the courts have been reviewed in the light of the state-of-the-art review. In studying these practical issues in this third Part it has turned out that some schools of science studies have been of significantly greater analytic value than others. For example, aside from the framing of the issue of dark matter itself, the claimed insights of Actor Network Theory have barely been invoked in making sense of risk or of science in public. Feminist science studies and the analysis of scientific discourse have not played a major role in the analyses in Chapters 8 to 12. Though actors' interests have been mentioned on many occasions, the formal apparatus of Interest Theory has not been deployed either. Insights from ethnomethodology have been referred to on a number of occasions though none of the detailed analyses has been undertaken from an ethnomethodological point of view. Still, even if the specific terminology of the different schools has played only a minor part, my principal conclusion is that the sensibility of science studies has been consistently important in making advances in the understanding of this series of substantive areas. As Bloor (1991) expressed it, the leading result of science studies is the 'finding' about the finitism of scientific judgement. I hope that I have been able to convince many sociologists that, on the basis of this apparently modest result and of the methodological commitments to symmetry and impartiality which gave rise to it, the sociology of science has a key role to play in identifying, exploring and illuminating society's dark matter. In my view, the sociology of science has to be acknowledged as a core component of social theory.

---

[1]In case this seems fanciful, see Woodcock and Davis, 1980: 112–15.

[2]Though, like Redner and Lyotard, I have repeatedly used the expression 'chaos theory' in the text, most mathematicians and physical scientists would now be reluctant to use this term and would class most aspects of chaos as a sub-set of the study of non-linear dynamics. 'Catastrophes' too would be regarded as a further form of non-linearity.

[3]The list of these in-between entities (fashionably known as cyborgs) is virtually limitless. Latour and Callon are also attracted to these cyborgs since, by straddling the human/non-human divide, they threaten to undermine the strong contrasts between human actors and the world of things which are pivotal to EPOR and most readings of the Strong Programme.

Bibliographical References

Agarwal, Anil and Narain, Sunita (1991) *Global Warming in an Unequal World: A Case of Environmental Colonialism*. Delhi: Centre for Science and Environment.

Amsterdamska, Olga (1990) 'Surely you are joking, Monsieur Latour!' [review of Latour, 1987], *Science, Technology and Human Values*, 15: 495–504.

Ashmore, Malcolm (1988) 'The life and opinions of a replication claim: reflexivity and symmetry in the sociology of scientific knowledge', in Steve Woolgar (ed.), *Knowledge and Reflexivity: New Frontiers in the Sociology of Knowledge*. London: Sage. pp. 125–53.

Ashmore, Malcolm (1989) *The Reflexive Thesis: Wrighting Sociology of Scientific Knowledge*. Chicago: University of Chicago Press.

Ashmore, Malcolm, Myers, Greg and Potter, Jonathan (1995) 'Discourse, rhetoric, reflexivity: seven days in the library', in Sheila Jasanoff, Gerald E. Markle, James C. Petersen and Trevor Pinch (eds), *Handbook of Science and Technology Studies*. London: Sage. pp. 321–42.

Barnes, Barry (1974) *Scientific Knowledge and Sociological Theory*. London: Routledge and Kegan Paul.

Barnes, Barry (1977) *Interests and the Growth of Knowledge*. London: Routledge and Kegan Paul.

Barnes, Barry and MacKenzie, Donald (1979) 'On the role of interests in scientific change', in Roy Wallis (ed.), *On the Margins of Science: The Social Construction of Rejected Knowledge, Sociological Review Monograph 27*. Keele: University of Keele. pp. 49–66.

Barnes, Barry, Bloor, David and Henry, John (1996) *Scientific Knowledge: A Sociological Analysis*. London: Athlone.

Beck, Ulrich (1992) *Risk Society: Towards a New Modernity*. London: Sage.

Beck, Ulrich (1995) *Ecological Politics in an Age of Risk*. Cambridge: Polity.

Begley, Sharon (2001) 'The science wars', in Muriel Lederman and Ingrid Bartsch (eds), *The Gender and Science Reader*. London: Routledge. pp. 114–18.

Bergström, Lars (1996) 'Scientific value', *International Studies in the Philosophy of Science*, 10: 189–202.

Bhaskar, Roy (1978) *A Realist Theory of Science*. Hassocks, Sussex: Harvester Press.

Bjelic, Dusan (1992) 'The praxiological validity of natural scientific practices as a criterion for identifying their unique social-object character: the case of the "authentication" of Goethe's morphological theorem', *Qualitative Sociology*, 15: 221–45.

Bjelic, Dusan and Lynch, Michael (1992) 'The work of a (scientific) demonstration: respecifying Newton's and Goethe's theories of prismatic colour', in Graham Watson and Robert M. Seiler (eds), *Text in Context: Contributions to Ethnomethodology*. London: Sage. pp. 52–78.

Black, Bert, Ayala, Francisco J. and Saffran-Brinks, Carol (1994) 'Science and the law in the wake of *Daubert*: a new search for scientific knowledge', *Texas Law Review*, 72: 715–802.

Bloor, David (1978) 'Polyhedra and the abominations of Leviticus', *British Journal for the History of Science*, 11: 243–72.

Bloor, David (1991) *Knowledge and Social Imagery* (2nd edition). Chicago: University of Chicago Press (original version 1976, London: Routledge and Kegan Paul).

Bodansky, Daniel (1994) 'Prologue to the climate change convention', in Irving M. Mintzer and J. Amber Leonard (eds), *Negotiating Climate Change: The Inside Story of the Rio Convention*. Cambridge: Cambridge University Press. pp. 45–74.

Boehmer-Christiansen, Sonja (1994a) 'Global climate protection policy: the limits of scientific advice, part 1', *Global Environmental Change*, 4: 140–59.

Boehmer-Christiansen, Sonja (1994b) 'Global climate protection policy: the limits of scientific advice, part 2', *Global Environmental Change*, 4: 185–200.

Boehmer-Christiansen, Sonja (2003) 'Science, equity, and the war against carbon', *Science, Technology and Human Values*, 28: 69–92.

Bowden, Gary (1985) 'The social construction of validity in estimates of US crude oil reserves', *Social Studies of Science*, 15: 207–40.

Brown, James R. (1989) *The Rational and the Social*. London: Routledge.

Bullard, Robert D. (1994) *Dumping in Dixie: Race, Class, and Environmental Quality*. Boulder, CO: Westview.

Burchfield, Joe D. (1990) *Lord Kelvin and the Age of the Earth*. Chicago: University of Chicago Press.

Button, Graham (1991) 'Introduction: ethnomethodology and the foundational respecification of the human sciences', in Graham Button (ed.), *Ethnomethodology and the Human Sciences*. Cambridge: Cambridge University Press. pp. 1–9.

Button, Graham and Sharrock, Wes (1993) 'A disagreement over agreement and consensus in constructionist sociology', *Journal for the Theory of Social Behaviour*, 23: 1–25.

Button, Graham and Sharrock, Wes (1995) 'The mundane work of writing and reading computer programs', in Paul ten Have and George Psathas (eds), *Situated Order: Studies in the Social Organization of Talk and Embodied Activities*. Washington, DC: International Institute for Ethnomethodology and Conversation Analysis and University Press of America. pp. 231–58.

Button, Graham and Sharrock, Wes (1998) 'The organizational accountability of technological work', *Social Studies of Science*, 28: 73–102.

Callon, Michel (1986) 'Some elements of a sociology of translation: domestication of the scallops and the fishermen of St Brieuc Bay', in John Law (ed.), *Power, Action and Belief: A New Sociology of Knowledge? Sociological*

Review Monograph 32. Keele: University of Keele. pp. 196–233 (reprinted in Mario Biagioli (ed.) (1999) *The Science Studies Reader*. London: Routledge).
Callon, Michel (1995) 'Four models for the dynamics of science', in Sheila Jasanoff, Gerald E. Markle, James C. Petersen and Trevor Pinch (eds), *Handbook of Science and Technology Studies*. London: Sage. pp. 29–63.
Callon, Michel and Latour, Bruno (1992) 'Don't throw the baby out with the Bath School! A reply to Collins and Yearley', in Andrew Pickering (ed.), *Science as Practice and Culture*. Chicago: University of Chicago Press. pp. 343–68.
Callon, Michel and Law, John (1982) 'On interests and their transformation: enrolment and counter-enrolment', *Social Studies of Science*, 12: 615–25.
Callon, Michel and Law, John (1997) 'Agency and the hybrid collectif', in Barbara Herrnstein Smith and Arkady Plotnitsky (eds), *Mathematics, Science and Postclassical Theory*. Durham, NC: Duke University Press.
Carey, John (ed.) (1995) *The Faber Book of Science*. London: Faber and Faber.
Collingridge, David and Reeve, Colin (1986) *Science Speaks to Power: The Role of Experts in Policymaking*. New York: St Martin's Press.
Collins, Harry M. (1981a) 'Stages in the empirical programme of relativism', *Social Studies of Science*, 11: 3–10.
Collins, Harry M. (1981b) 'What is TRASP?: The radical programme as a methodological imperative', *Philosophy of the Social Sciences*, 11: 215–24.
Collins, Harry M. (1983) 'An empirical relativist programme in the sociology of scientific knowledge', in Karin D. Knorr-Cetina and Michael Mulkay (eds), *Science Observed: Perspectives on the Social Study of Science*. London: Sage. pp. 85–113.
Collins, Harry M. (1992) *Changing Order: Replication and Induction in Scientific Practice*. Chicago: University of Chicago Press.
Collins, Harry M. (1996) 'In praise of futile gestures: how scientific is the sociology of scientific knowledge?', *Social Studies of Science*, 26: 229–44.
Collins, Harry M. (1998) 'The meaning of data: open and closed evidential cultures in the search for gravitational waves', *American Journal of Sociology*, 104: 293–338.
Collins, Harry M. and Pinch, Trevor J. (1993) *The Golem: What Everyone Should Know about Science*. Cambridge: Cambridge University Press.
Collins, Harry M. and Yearley, Steven (1992a) 'Epistemological chicken', in Andrew Pickering (ed.), *Science as Practice and Culture*. Chicago: University of Chicago Press. pp. 301–26.
Collins, Harry M. and Yearley, Steven (1992b) 'Journey into space', in Andrew Pickering (ed.), *Science as Practice and Culture*. Chicago: University of Chicago Press. pp. 369–89.
Dean, John (1979) 'Controversy over classification: a case study from the history of botany', in Barry Barnes and Steven Shapin (eds), *Natural Order: Historical Studies of Scientific Culture*. London: Sage. pp. 211–30.
Delamont, Sara (1987) 'Three blind spots? A comment on the sociology of science by a puzzled outsider', *Social Studies of Science*, 17: 163–70.
Dennis, Michael A. (1985) 'Drilling for dollars: the making of US petroleum reserve estimates, 1921–25', *Social Studies of Science*, 15: 241–65.

Douglas, Mary and Wildavsky, Aaron (1982) *Risk and Culture: An Essay on the Selection of Technological and Environmental Dangers*. Berkeley, CA: University of California Press.

Dowie, Mark (1995) *Losing Ground: American Environmentalism at the Close of the Twentieth Century*. London: MIT Press.

Dugan K.G. (1987) 'The zoological exploration of the Australian region and its impact on biological theory', in Nathan Reingold and Marc Rothenberg (eds), *Scientific Colonialism: A Cross-Cultural Comparison*. Washington DC: Smithsonian Institution Press. pp. 79–100.

Durant, John R., Evans, Geoffrey A. and Thomas, Geoffrey P. (1989) 'The public understanding of science', *Nature*, 340: 6 July. pp. 11–14.

Durant, John R., Evans, Geoffrey and Thomas, Geoffrey (1992) 'Public understanding of science in Britain: the role of medicine in the popular representation of science', *Public Understanding of Science*, 1: 161–82.

Edmond, Gary (2002) 'Legal engineering: contested representations of law, science (and non-science) and society', *Social Studies of Science*, 32: 371–412.

Edmond, Gary and Mercer, David (1999) 'Creating (public) science in the *Noah's Ark case*', *Public Understanding of Science*, 8: 317–43.

Edmond, Gary and Mercer, David (2000) Litigation life: law-science knowledge construction in (Bendectin) mass toxic tort litigation', *Social Studies of Science*, 30: 265–316.

Epstein, Steven (1995) 'The construction of lay expertise: AIDS activism and the forging of credibility in the reform of clinical trials', *Science, Technology and Human Values*, 15: 495–504.

Evans, Geoffrey and Durant, John (1995) 'The relationship between knowledge and attitudes in the public understanding of science in Britain', *Public Understanding of Science*, 4: 57–74.

Farley, John and Geison, Gerald L. (1982) 'Science, politics and spontaneous generation in nineteenth-century France: the Pasteur–Pouchet debate', in Harry M. Collins (ed.), *Sociology of Scientific Knowledge: A Source Book*. Bath: Bath University Press. pp. 1–38.

Fischoff, Baruch, Slovic, Paul, Lichtenstein, Sarah, Read, S. and Combs, B. (1978) 'How safe is safe enough? A psychometric study of attitudes towards technological risks and benefits', *Policy Sciences*, 9: 127–52.

Foster, Kenneth R. and Huber, Peter W. (1997) *Judging Science: Scientific Knowledge and the Federal Courts*. London: MIT Press.

Galison, Peter (1987) *How Experiments End*. Chicago: University of Chicago Press.

Garfinkel, Harold (1967) *Studies in Ethnomethodology*. Englewood Cliffs, NJ: Prentice-Hall.

Garfinkel, Harold (1996) 'Ethnomethodology's program', *Social Psychology Quarterly*, 59: 5–21.

Garfinkel, Harold, Lynch, Michael and Livingston, Eric (1981) 'The work of a discovering science construed with materials from the optically discovered pulsar', *Philosophy of the Social Sciences*, 11: 131–58.

Gibbons, Michael, Limoges, Camille, Nowotny, Helga, Schwartzman, Simon, Scott, Peter and Trow, Martin (1994) *The New Production of Knowledge*. London: Sage.

Giddens, Anthony (2002) *Runaway World*. London: Profile Books.
Gieryn, Thomas F. (1999) *Cultural Boundaries of Science: Credibility on the Line*. Chicago: University of Chicago Press.
Gilbert, G. Nigel and Mulkay, Michael (1980) 'Contexts of scientific discourse: social accounting in experimental papers', in Karin D. Knorr, Roger Krohn and Richard Whitley (eds), *The Social Process of Scientific Investigation, Sociology of the Sciences Yearbook IV*. Dordrecht: Reidel. pp. 269–94.
Gilbert, G. Nigel and Mulkay, Michael (1984) *Opening Pandora's Box*. Cambridge: Cambridge University Press.
Gross, Paul R. and Levitt, Norman (1994) *Higher Superstition: The Academic Left and its Quarrels with Science*. Baltimore, MD: Johns Hopkins University Press.
Haas, Peter M. (1992) 'Introduction: epistemic communities and international policy coordination', *International Organization*, 46: 1–35.
Habermas, Jürgen (1971) *Toward a Rational Society: Student Protest, Science and Politics*. London: Heinemann.
Habermas, Jürgen (1972) *Knowledge and Human Interests*. London: Heinemann.
Habermas, Jürgen (1973) 'A postscript to *Knowledge and Human Interests*', *Philosophy of the Social Sciences*, 3: 157–89.
Hacking, Ian (1990) *The Taming of Chance*. Cambridge: Cambridge University Press.
Hacking, Ian (1999) *The Social Construction of What?* Cambridge, MA: Harvard University Press.
Hammond, A.L., Rodenburg, E. and Moomaw, W.R. (1991) 'Calculating national accountability for climate change', *Environment*, 33: 11–35.
Harding, Sandra (1986) *The Science Question in Feminism*. Milton Keynes: Open University Press.
Harding, Sandra (1991) *Whose Science? Whose Knowledge? Thinking from Women's Lives*. Milton Keynes: Open University Press.
Hartsock, Nancy C.M. (1983) 'The feminist standpoint: developing the ground for a specifically feminist historical materialism', in Sandra Harding and Merrill B. Hintikka (eds), *Discovering Reality: Feminist Perspectives on Epistemology, Metaphysics, Methodology, and Philosophy of Science*. Dordrecht: Reidel. pp. 283–310.
Heritage, John (1984) *Garfinkel and Ethnomethodology*. Cambridge: Polity.
Hubbard, Ruth (1990) *The Politics of Women's Biology*. New Brunswick, NJ: Rutgers University Press.
Irwin, Alan and Wynne, Brian (1996) 'Introduction', in Alan Irwin and Brian Wynne (eds), *Misunderstanding Science? The Public Reconstruction of Science and Technology*. Cambridge: Cambridge University Press. pp. 1–17.
Irwin, Alan, Dale, Alison and Smith, Denis (1996) 'Science and Hell's kitchen: the local understanding of hazard issues', in Alan Irwin and Brian Wynne (eds), *Misunderstanding Science? The Public Reconstruction of Science and Technology*. Cambridge: Cambridge University Press. pp. 47–64.
Irwin, Alan, Rothstein, Henry, Yearley, Steven and McCarthy, Elaine (1997) 'Regulatory science – towards a sociological framework', *Futures*, 29: 17–31.
Jasanoff, Sheila S. (1986) *Risk Management and Political Culture*. New York: Russell Sage.

Jasanoff, Sheila S. (1990) *The Fifth Branch: Science Advisers as Policymakers*. Cambridge, MA: Harvard University Press.

Jasanoff, Sheila S. (1995) *Science at the Bar: Law, Science, and Technology in America*. Cambridge, MA: Harvard University Press.

Jasanoff, Sheila S. (1996) 'Science and norms in global environmental regimes', in Fen O. Hampson and Judith Reppy (eds), *Earthly Goods: Environmental Change and Social Justice*. Ithaca, NY: Cornell University Press. pp. 173–97.

Jasanoff, Sheila S. (1997) 'Civilization and Madness: the great BSE scare of 1996', *Public Understanding of Science*, 6: 221–32.

Jasanoff, Sheila S. (1998) 'The eye of everyman: witnessing DNA in the Simpson trial', *Social Studies of Science*, 28: 713–40.

Jasanoff, Sheila S. (1999) 'The Songlines of risk', *Environmental Values*, 8: 135–52.

Jasanoff, Sheila S. (2001) 'Hidden experts: judging science after *Daubert*', in Vivian Weil (ed.), *Trying Times: Science and Responsibilities after Daubert*. Chicago: Illinois Institute of Technology. pp. 30–47.

Jasanoff, Sheila S. and Wynne, Brian (1998) 'Science and decisionmaking', in Steve Rayner and Elizabeth L. Malone (eds), *Human Choice and Climate Change*, Vol. 1. Columbus, OH: Battelle Press. pp. 1–87.

Kerr, E. Anne (2001) 'Toward a feminist natural science: linking theory and practice', in Muriel Lederman and Ingrid Bartsch (eds), *The Gender and Science Reader*. London: Routledge. pp. 386–406.

Kitcher, Philip (1993) *The Advancement of Science*. New York: Oxford University Press.

Kitcher, Philip (1996) *The Lives to Come: The Genetic Revolution and Human Possibilities*. New York: Simon and Schuster.

Knorr-Cetina, Karin D. (1981) *The Manufacture of Knowledge*. Oxford: Pergamon.

Kuhn, Thomas S. (1970) 'Logic of discovery or psychology of research', in Imre Lakatos and Alan Musgrave (eds), *Criticism and the Growth of Knowledge*. Cambridge: Cambridge University Press. pp. 1–23.

Kuhn, Thomas S. (1977) *The Essential Tension: Selected Studies in Scientific Tradition and Change*. Chicago: University of Chicago Press.

Lakatos, Imre (1978) *The Methodology of Scientific Research Programmes*. Cambridge: Cambridge University Press.

Lambert, Helen and Rose, Hilary (1996) 'Disembodied knowledge? Making sense of medical science', in Alan Irwin and Brian Wynne (eds), *Misunderstanding Science? The Public Reconstruction of Science and Technology*. Cambridge: Cambridge University Press. pp. 65–83.

Latour, Bruno (1983) 'Give me a laboratory and I will raise the world', in Karin D. Knorr-Cetina and Michael Mulkay (eds), *Science Observed: Perspectives on the Social Study of Science*. London: Sage. pp. 141–70.

Latour, Bruno (1984) 'Where did you put the black-box opener?' [review of Gilbert and Mulkay 1984], *EASST Newsletter*, 3: 17–21.

Latour, Bruno (1987) *Science in Action*. Milton Keynes: Open University Press.

Latour, Bruno (1988a) *The Pasteurization of France*. Cambridge, MA: Harvard University Press.

Latour, Bruno (1988b) 'The politics of explanation: an alternative', in Steve Woolgar (ed.), *Knowledge and Reflexivity: New Frontiers in the Sociology of Knowledge*. London: Sage. pp. 155–76.
Latour, Bruno (1992) 'Where are the missing masses? The sociology of a few mundane artifacts', in Wiebe E. Bijker and John Law (eds), *Shaping Technology/ Building Society: Studies in Sociotechnical Change*. London: MIT Press. pp. 225–58.
Latour, Bruno (1993) *We Have Never Been Modern*. Hemel Hempstead and London: Harvester Wheatsheaf.
Latour, Bruno (1999a) 'On recalling ANT', in John Law and John Hassard (eds), *Actor Network Theory and After*. Oxford: Blackwell. pp. 15–25.
Latour, Bruno (1999b) *Pandora's Hope: Essays on the Reality of Science Studies*. Cambridge, MA: Harvard University Press.
Latour, Bruno (2000) 'When things strike back: a possible contribution of "science studies" to the social sciences', *British Journal of Sociology*, 51: 107–23.
Latour, Bruno and Woolgar, Steve (1979) *Laboratory Life: The Social Construction of Scientific Facts*. London: Sage.
Laudan, Larry (1977) *Progress and its Problems: Towards a Theory of Scientific Growth*. Berkeley, CA: University of California Press.
Laudan, Larry (1982) 'A note on Collins' blend of relativism and empiricism', *Social Studies of Science*, 12: 131–2.
Livingston, Eric (1999) 'Cultures of proving', *Social Studies of Science*, 29: 867–88.
Longino, Helen E. (1989) 'Can there be a feminist science?' in Ann Garry and Marilyn Pearsall (eds), *Women, Knowledge and Reality: Explorations in Feminist Philosophy*. London: Unwin Hyman. pp. 203–16.
Longino, Helen E. (1990) *Science as Social Knowledge*. Princeton, NJ: Princeton University Press.
Longino, Helen E. (2002) *The Fate of Knowledge*. Princeton, NJ: Princeton University Press.
Lowe, Philip, Clark, Judy, Seymour, Susanne and Ward, Neil (1997) *Moralizing the Environment: Countryside Change, Farming and Pollution*. London: UCL Press.
Lukács, Georg (1971) *History and Class Consciousness*. London: Merlin.
Lynch, Michael (1985) *Art and Artifact in Laboratory Science: A Study of Shop Work and Shop Talk in a Research Laboratory*. London: Routledge and Kegan Paul.
Lynch, Michael (1992) 'Extending Wittgenstein: the pivotal move from epistemology to the sociology of science', in Andrew Pickering (ed.), *Science as Practice and Culture*. Chicago: University of Chicago Press. pp. 343–68.
Lynch, Michael (1993) *Scientific Practice and Ordinary Action: Ethnomethodology and Social Studies of Science*. Cambridge: Cambridge University Press.
Lynch, Michael (2001) 'A pragmatogony of factishes', [review of Latour, 1999b], *Metascience: An International Journal for the History, Philosophy and Social Studies of Science*, 10: 223–32.
Lynch, Michael (2002) 'Protocols, practices, and the reproduction of technique in molecular biology', *British Journal of Sociology*, 53: 203–20.
Lynch, Michael and Jasanoff, Sheila S. (1998) 'Contested identities: science, law and forensic practice', *Social Studies of Science*, 28: 675–86.

Lynch, Michael, Livingston, Eric and Garfinkel, Harold (1983) 'Temporal order in laboratory work', in Karin D. Knorr-Cetina and Michael Mulkay (eds), *Science Observed: Perspectives on the Social Study of Science*. London: Sage. pp. 205–38.

Lyotard, Jean-François (1984) *The Postmodern Condition: A Report on Knowledge*. Manchester: Manchester University Press.

MacKenzie, Donald (1978) 'Statistical theory and social interests: a case-study', *Social Studies of Science*, 8: 35–83.

MacKenzie, Donald (1981) *Statistics in Britain, 1865–1930: The Social Construction of Scientific Knowledge*. Edinburgh: Edinburgh University Press.

MacKenzie, Donald (1984) 'Reply to Steven Yearley', *Studies in History and Philosophy of Science*, 15: 251–9.

MacKenzie, Donald and Barnes, Barry (1979) 'Scientific judgment: the Biometry–Mendelism controversy', in Barry Barnes and Steven Shapin (eds), *Natural Order: Historical Studies of Scientific Culture*. London: Sage. pp. 191–210.

Martin, Emily (1996) 'The egg and the sperm: how science has constructed a romance based on stereotypical male–female roles', in Barbara Laslett, Sally Gregory Kohlstedt, Helen Longino and Evelynn Hammonds (eds), *Gender and Scientific Authority*. Chicago: Chicago University Press. pp. 323–39 (reprinted in Stevi Jackson and Sue Scott (eds), (2001) *Gender: A Sociological Reader*. London: Routledge).

May, Robert (1992) 'The chaotic rhythms of life', in Nina Hall (ed.), *The New Scientist Guide to Chaos*. London: Penguin. pp. 82–95.

McCright, Aaron M. and Dunlap, Riley E. (2000) 'Challenging global warming as a social problem: an analysis of the conservative movement's counter-claims', *Social Problems*, 47: 499–522.

McKinlay, Andrew and Potter, Jonathan (1987) 'Model discourse: interpretative repertoires in scientists' conference talk', *Social Studies of Science*, 17: 443–63.

Merton, Robert K. (1973) *The Sociology of Science: Theoretical and Empirical Investigations*. Chicago: Chicago University Press.

Mitroff, Ian I. (1974) *The Subjective Side of Science*. New York: Elsevier.

Mulkay, Michael (1976) 'Norms and ideology in science', *Social Science Information*, 15: 637–56.

Mulkay, Michael (1980) 'Interpretation and the use of rules: the case of the norms of science', in Thomas F. Gieryn (ed.), *Science and Social Structure: A Festschrift for Robert K. Merton, Transactions of the New York Academy of Sciences Series II, Volume 39*. New York: Academy of Sciences. pp. 111–25.

Mulkay, Michael (1981) 'Action and belief or scientific discourse? A possible way of ending intellectual vassalage in social studies of science', *Philosophy of the Social Sciences*, 11: 163–71.

Mulkay, Michael (1984) 'The ultimate compliment: a sociological analysis of ceremonial discourse', *Sociology*, 18: 531–49.

Mulkay, Michael (1988) 'Don Quixote's double: a self-exemplifying text', in Steve Woolgar (ed.), *Knowledge and Reflexivity: New Frontiers in the Sociology of Knowledge*. London: Sage. pp. 81–100.

Mulkay, Michael (1991) *Sociology of Science: A Sociological Pilgrimage*. Milton Keynes: Open University Press.

Mulkay, Michael (1993) 'Rhetorics of hope and fear in the great embryo debate', *Social Studies of Science*, 23: 721–42.

Mulkay, Michael and Gilbert, G. Nigel (1981) 'Putting philosophy to work: Karl Popper's influence on scientific practice', *Philosophy of the Social Sciences*, 11: 389–407.

Mulkay, Michael and Gilbert, G. Nigel (1982a) 'Accounting for error: how scientists construct their social world when they account for correct and incorrect belief', *Sociology*, 16: 165–83.

Mulkay, Michael and Gilbert, G. Nigel (1982b) 'Joking apart: some recommendations concerning the analysis of scientific culture', *Social Studies of Science*, 12: 585–613.

Nelkin, Dorothy and Lindee, M. Susan (1995) *The DNA Mystique: The Gene as a Cultural Icon*. New York: W.H. Freeman.

Newton-Smith, William H. (1981) *The Rationality of Science*. London: Routledge and Kegan Paul.

Nowotny, Helga, Scott, Peter and Gibbons, Michael (2001) *Re-Thinking Science: Knowledge and the Public in an Age of Uncertainty*. Cambridge: Polity.

Oakley, Ann (1972) *Sex, Gender and Society*. London: Temple Smith.

Oteri, J.S., Weinberg, M.G. and Pinales, M.S. (1982) 'Cross-examination of chemists in drug cases', in Barry Barnes and David Edge (eds), *Science in Context: Readings in the Sociology of Science*. Milton Keynes: Open University Press. pp. 250–9.

Palmer, Derrol (2000) 'Identifying delusional discourse: issues of rationality, reality and power', *Sociology of Health and Illness*, 22: 661–78.

Palmer, Tim (1992) 'A weather eye on unpredictability', in Nina Hall (ed.), *The New Scientist Guide to Chaos*. London: Penguin. pp. 69–81.

Pearce, Fred (1991) *Green Warriors: The People and the Politics Behind the Environmental Revolution*. London: Bodley Head.

Pels, Dick (1996) 'The politics of symmetry', *Social Studies of Science*, 26: 277–304.

Pickering, Andrew (1980) 'The role of interests in high-energy physics: the choice between charm and colour', in Karin D. Knorr, Roger Krohn and Richard Whitley (eds), *The Social Process of Scientific Investigation*. Dordrecht: Reidel. pp. 107–38.

Pickering, Andrew (1984) *Constructing Quarks: a Sociological History of Particle Physics*. Edinburgh: Edinburgh University Press.

Pickering, Andrew (1995) *The Mangle of Practice: Time, Agency and Science*. Chicago: University of Chicago Press.

Pickering, Kevin T. and Owen, Lewis A. (1994) *An Introduction to Global Environmental Issues*. London: Routledge.

Pinch, Trevor (1980) 'Theoreticians and the production of experimental anomaly: the case of solar neutrinos', in Karin D. Knorr, Roger Krohn and Richard Whitley (eds), *The Social Process of Scientific Investigation*. Dordrecht: Reidel. pp. 77–106.

Pinch, Trevor (1993) 'Generations of SSK' [review of Richards 1991], *Social Studies of Science*, 23: 363–73.

Pinch, Trevor and Pinch, Trevor (1988) 'Reservations about reflexivity and New Literary Forms or why let the devil have all the good tunes?', in Steve

Woolgar (ed.), *Knowledge and Reflexivity: New Frontiers in the Sociology of Knowledge*. London: Sage. pp. 178–97.

Poovey, Mary (1998) *A History of the Modern Fact: Problems of Knowledge in the Sciences of Wealth and Society*. Chicago: University of Chicago Press.

Popper, Karl R. (1972a) *Conjectures and Refutations: The Growth of Scientific Knowledge*. London: Routledge and Kegan Paul.

Popper, Karl R. (1972b) *Objective Knowledge: An Evolutionary Approach*. Oxford: Oxford University Press.

Porter, Theodore M. (1986) *The Rise of Statistical Thinking 1820–1900*. Princeton, NJ: Princeton University Press.

Putnam, Hilary (1994) 'Sense, nonsense, and the senses: an inquiry into the powers of the human mind', *Journal of Philosophy*, 91: 445–517.

Quine, Willard Van Orman (1969) *Ontological Relativity and Other Essays*. New York: Columbia University Press.

Redner, Harry (1987) *The Ends of Science*. Boulder, CO: Westview.

Redner, Harry (1994) *A New Science of Representation: Towards an Integrated Theory of Representation in Science, Politics and Art*. Boulder, CO: Westview.

Richards, Evelleen (1991) *Vitamin C and Cancer: Medicine or Politics?*. London: Macmillan.

Richards, Evelleen (1996) '(Un)boxing the monster', *Social Studies of Science*, 26: 323–56.

Rip, Arie (1982) 'De gans met de gouden eieren en andere maatschappelijke legitimaties van de moderne wetenschap', *De Gids*, 145: 285–97.

Schaffer, Simon (1991) 'The Eighteenth Brumaire of Bruno Latour', *Studies in History and Philosophy of Science*, 22: 174–92.

Schiebinger, Londa (1999) *Has Feminism Changed Science?* Cambridge, MA: Harvard University Press.

Scott, Pam, Richards, Evelleen and Martin, Brian (1990) 'Captives of controversy: the myth of the neutral social researcher in contemporary scientific controversies', *Science, Technology and Human Values*, 15: 474–94.

Shapin, Steven (1979) 'The politics of observation: cerebral anatomy and social interests in the Edinburgh phrenology disputes', in Roy Wallis (ed.), *On the Margins of Science: The Social Construction of Rejected Knowledge, Sociological Review Monograph 27*. Keele: University of Keele. pp. 139–78.

Shapin, Steven (1994) *A Social History of Truth: Civility and Science in Seventeenth-Century England*. Chicago: University of Chicago Press.

Shapin, Steven (1995) 'Here and everywhere: sociology of scientific knowledge', *Annual Review of Sociology*, 21: 289–321.

Sharrock, Wes and Anderson, Bob (1991) 'Epistemology: professional scepticism', in Graham Button (ed.), *Ethnomethodology and the Human Sciences*. Cambridge: Cambridge University Press. pp. 51–76.

Simmons, Ian G. (1993) *Interpreting Nature: Cultural Constructions of the Environment*. London: Routledge.

Slovic, Paul (1992) 'Perception of risk: reflections on the psychometric paradigm', in Sheldon Krimsky and Dominic Golding (eds), *Social Theories of Risk*. London: Praeger. pp. 117–52.

Solomon, Shana M. and Hackett, Edward J. (1996) 'Setting boundaries between science and the law: lessons from *Daubert v. Merrell Dow Pharmaceuticals, Inc.*', *Science, Technology and Human Values*, 21: 131–56.

Star, Susan Leigh and Griesemer, James R. (1989) 'Institutional ecology, "translations" and boundary objects: amateurs and professionals in Berkeley's Museum of Vertebrate Zoology, 1907–39', *Social Studies of Science*, 19: 387–420.

Stirling, Andy and Mayer, Sue (1999) *Re-Thinking Risk: A Pilot Multi-Criteria Mapping of a Genetically Modified Crop in Agricultural Systems in the UK*. Brighton, Sussex: Science Policy Research Unit.

Stoppard, Tom (1999) *Plays 5*. London: Faber and Faber.

Tuana, Nancy (1989) 'Preface' in Nancy Tuana (ed.), *Feminism and Science*. Bloomington, IN: Indiana University Press. pp. vii–xi.

Turner, Stephen (2001) 'What is the problem with experts?', *Social Studies of Science*, 31: 123–49.

UK Government (1993) *Realising our Potential: A Strategy for Science, Engineering and Technology*. London: HMSO (Cm 2250).

Ward, Steven C. (1996) *Reconfiguring Truth: Postmodernism, Science Studies and the Search for a New Model of Knowledge*. London: Rowman and Littlefield.

Warner, Frederick (1992) *Risk: Analysis, Perception and Management: Report of a Royal Society Study Group*. London: Royal Society.

Weinberg, Alvin M. (1972) 'Science and trans-science', *Minerva*, 10: 209–22.

Whitley, Richard (1984) *The Intellectual and Social Organization of the Sciences*. Oxford: Clarendon Press.

Woodcock, Alexander and Davis, Monte (1980) *Catastrophe Theory*. Harmondsworth: Penguin.

Woolgar, Steve (1981) 'Interests and explanation in the social study of science', *Social Studies of Science*, 11: 365–94.

Woolgar, Steve (1983) 'Irony in the social study of science', in Karin D. Knorr-Cetina and Michael Mulkay (eds), *Science Observed: Perspectives on the Social Study of Science*. London: Sage. pp. 239–66.

Woolgar, Steve and Ashmore, Malcolm (1988) 'The next step: an introduction to the reflexive project', in Steve Woolgar (ed.), *Knowledge and Reflexivity: New Frontiers in the Sociology of Knowledge*. London: Sage. pp. 1–11.

World Resources Institute (1990) *World Resources 1990–91*. New York: Oxford University Press.

Wynne, Brian (1982) *Rationality and Ritual: The Windscale Inquiry and Nuclear Decisions in Britain*. Chalfont St Giles: British Society for the History of Science.

Wynne, Brian (1989) 'Frameworks of rationality in risk management', in J. Brown (ed.), *Environmental Threats: Perception, Analysis and Management*. London: Belhaven. pp. 33–47.

Wynne, Brian (1992a) 'Misunderstood misunderstanding: social identities and public uptake of science', *Public Understanding of Science*, 1: 281–304.

Wynne, Brian (1992b) 'Uncertainty and environmental learning', *Global Environmental Change*, 2: 111–27.

Wynne, Brian (1995) 'Public understanding of science', in Sheila Jasanoff, Gerald E. Markle, James C. Petersen and Trevor Pinch (eds), *Handbook of Science and Technology Studies*. London: Sage. pp. 361–88.

Wynne, Brian (1996) 'May the sheep safely graze? A reflexive view of the expert–lay knowledge divide', in Scott Lash, Bronislaw Szerszynski and Brian Wynne (eds), *Risk, Environment and Modernity: Towards a New Ecology*. London: Sage. pp. 27–43.

Wynne, Brian (2001) 'Creating public alienation: expert cultures of risk and ethics on GMOs', *Science as Culture*, 10: 445–81.

Yearley, Steven (1981) 'Textual persuasion: the role of social accounting in the construction of scientific arguments', *Philosophy of the Social Sciences*, 11: 409–35.

Yearley, Steven (1982) 'The relationship between epistemological and sociological cognitive interests', *Studies in History and Philosophy of Science*, 13: 353–88.

Yearley, Steven (1984) *Science and Sociological Practice*. Milton Keynes: Open University Press.

Yearley, Steven (1987) 'The two faces of science' [review of Latour, 1987], *Nature*, 326 (23–29 April): 754.

Yearley, Steven (1989) 'Bog standards: science and conservation at a public inquiry', *Social Studies of Science*, 19: 421–38.

Yearley, Steven (1992) 'Skills, deals and impartiality: the sale of environmental consultancy skills and public perceptions of scientific neutrality', *Social Studies of Science*, 22: 435–53.

Yearley, Steven (1993) [review of Richards, 1991], *Social History of Medicine*, 6: 299–300.

Yearley, Steven (1995) 'Environmental attitudes in Northern Ireland', in Richard Breen, Paula Devine and Gillian Robinson (eds), *Social Attitudes in Northern Ireland: The Fourth Report 1994–1995*. Belfast: Appletree Press. pp. 119–41.

Yearley, Steven (1996) *Sociology, Environmentalism, Globalization*. London: Sage.

Yearley, Steven (1997) 'The changing social authority of science', *Science Studies*, 1: 65–75.

Yearley, Steven (1999a) 'Computer models and the public's understanding of science: a case-study analysis', *Social Studies of Science*, 29: 845–66.

Yearley, Steven (1999b) [review of Lowe et al., 1997], *Environmental Policy*, 8: 184–5.

Yearley, Steven (2000) 'Making systematic sense of public discontents with expert knowledge: two analytical approaches and a case study', *Public Understanding of Science*, 9: 105–22.

Zahar, Elie (1973) 'Why did Einstein's programme supersede Lorentz's?', *British Journal for the Philosophy of Science*, 24: 95–123.

Zehr, Stephen C. (2000) 'Public representations of scientific uncertainty about global climate change', *Public Understanding of Science*, 9: 85–103.

Ziman, John M. (1978) *Reliable Knowledge: An Exploration of the Grounds for Belief in Science*. Cambridge: Cambridge University Press.

Index